Geoheritage, Geoparks and Geotourism

Conservation and Management Series

Series Editors

Wolfgang Eder, GeoCentre-Geobiology, University of Göttingen, Göttingen, Niedersachsen, Germany

Peter T. Bobrowsky, Geological Survey of Canada, Sidney, BC, Canada

Jesús Martínez-Frías, CSIC-Universidad Complutense de Madrid, Instituto de Geociencias, Madrid, Spain

Spectacular geo-morphological landscapes and regions with special geological features or mining sites are becoming increasingly recognized as critical areas to protect and conserve for the unique geoscientific aspects they represent and as places to enjoy and learn about the science and history of our planet. More and more national and international stakeholders are engaged in projects related to "Geoheritage", "Geo-conservation", "Geoparks" and "Geo-tourism"; and are positively influencing the general perception of modern Earth Sciences. Most notably, "Geoparks" have proven to be excellent tools to educate the public about Earth Sciences; and they are also important areas for recreation and significant sustainable economic development through geotourism. In order to develop further the understanding of Earth Sciences in general and to elucidate the importance of Earth Sciences for Society, the "Geoheritage, Geoparks and Geotourism Conservation and Management Series" has been launched together with its sister "GeoGuides" series. Projects developed in partnership with UNESCO, World Heritage and Global Geoparks Networks, IUGS and IGU, as well as with the 'Earth Science Matters' Foundation will be considered for publication. This series aims to provide a place for in-depth presentations of developmental and management issues related to Geoheritage and Geotourism in existing and potential Geoparks. Individually authored monographs as well as edited volumes and conference proceedings are welcome; and this book series is considered to be complementary to the Springer-Journal "Geoheritage".

More information about this series at http://www.springer.com/series/11639

Irakli Gamkrelidze • Avtandil Okrostsvaridze • Kakhaber Koiava • Ferando Maisadze

Geotourism Potential of Georgia, the Caucasus

History, Culture, Geology, Geotourist Routes and Geoparks

Irakli Gamkrelidze
Aleksandre Janelidze Institute of Geology
Tbilisi State University (TSU)
Tbilisi, Georgia

Kakhaber Koiava
Aleksandre Janelidze Institute of Geology
Tbilisi State University (TSU)
Tbilisi, Georgia

Avtandil Okrostsvaridze
Ilia State University
Tbilisi, Georgia

Ferando Maisadze
Aleksandre Janelidze Institute of Geology
Tbilisi State University (TSU)
Tbilisi, Georgia

ISSN 2363-765X   ISSN 2363-7668   (electronic)
Geoheritage, Geoparks and Geotourism
ISBN 978-3-030-62968-7   ISBN 978-3-030-62966-3   (eBook)
https://doi.org/10.1007/978-3-030-62966-3

This Springer imprint is published by the registered company Springer Nature Switzerland AG
The registered company address is: Gewerbestrasse 11, 6330 Cham, Switzerland

*We dedicate the presented work to
The 95th anniversary of the
establishment of Aleksandre Janelidze Institute
of Geology, founded in 1925
at the Tbilisi State University by the
leadership of academician Aleksandre
Janelidze—the founder of the
Georgian geological school.*

Irakli Gamkrelidze

Avtandil Okrostsvaridze

Kakhaber Koiava

Ferando Maisadze

Preface

Georgia is a country with an ancient history, rich culture, various geographical areas and good geological outcrops. The book provides a more extensive geological and brief geographic, historical and cultural overview of this country and is aimed at increasing and developing the country's geotourism potential and joining to UNESCO Global Geopark Network.

Georgia is located in the mountain area, which is known as the Caucasus from ancient times. It borders Russia to the north along the Greater Caucasus Ridge, Azerbaijan to the east, Armenia to the south and Turkey to the southwest. Its west margin borders the Black Sea. Georgia is 69,500 km^2 in area; its population is nearly 4 million. Representatives of more than 10 nations and nationalities live in it. About 86.8% of the population are Georgians.

The territory of Georgia includes the Transcaucasian intermountain area, the southern slope of the Greater Caucasus and a big part of the Lesser Caucasus. Because of its accessible terrain and good climate, the Caucasus intermountain depression is the most favorable natural corridor between Black and Caspian Seas, as well as Europe and Asia. It is protected from cold air masses invasion by the Greater Caucasus Range from the north, and hot and dry air masses— by the Lesser Caucasus from the south. This segment of the Caucasus is notable for its favorable climatic conditions for development of bio-populations, including humans, which is the reason that the area has been inhabited by early Hominids since Early Pleistocene (Dmanisi Hominids site −1.81 Ma). In addition, numerous Paleolithic age humans' occupations are discovered here, which means that the area has been widely inhabited by people from the Late Pleistocene.

We, the authors of this book all our life study the geological structure of Georgia, the Caucasus and adjacent areas of the Alpine-Himalayan orogenic belt. For many years, we introduced our foreign colleagues to the unique geology of Georgia, which determines the presence here of very interesting geotourism objects—geotourist routes and potential geoparks.

Because of the small area, it is possible to become familiar in Georgia with the whole spectrum of rocks during several days. For example, only one day is enough to see both Precambrian relicts (Dzirula massif) and Quaternary volcanoes (Abul-Samsari Range). Also, during a day, one can get to know the whole cross section of the collisional zone starting from the Fold system of the Lesser Caucasus (from Tbilisi) and ending with megastructure of the Greater Caucasus (Kazbegi region).

Georgia provides an opportunity to see the Greater Caucasus glacier-covered alpine gorgeous mountain peaks; footprints of Cretaceous dinosaurs at Sataplia; Borjomi and Tskaltubo mineral water deposits; beautiful landscape of Gareja semi-desert, built up of Miocene sedimentary layers; glacier-capped fascinating Quaternary volcano Kazbegi and many other interesting geological objects. But the diversity of geologic objects is not the only reason to be fascinated by Georgia. It is also home to significant paleoarchaeological, archaeological and historic sites that can be visited concurrently during the geological tours.

Acknowledgements We would like to express our gratitude to our colleague Dr. Eteri Kilasonia for technical assistance and Lika Kvaliashvili for linguistic corrections.

We are very grateful to the staff at Springer, in particular Dr. Alexis Vizcaino and Karthik Raj Selvaraj, for their help and support. We also would like to thank the reviewers for their constructive suggestions.

Irakli Gamkrelidze
Avtandil Okrostsvaridze
Kakhaber Koiava
Ferando Maisadze

Contents

Overview of the History and Culture of Georgia

Georgia is a country with ancient history and culture. The territory of Georgia has been a favorable area for living since ancient times. This is confirmed by archeological excavations in Dmanisi. Fragments of human skeletons found in Dmanisi are the oldest in whole Eurasia and are between 1,600,000 and 1,800,000 years old. Older skeletons are found in Africa. It can be said that the people of Dmanisi are the ancient inhabitants of Eurasia (Gabunia et al. 2000).

Georgian tribes appeared in Western Asia in thirteenth century BC, when two ancient states—the Hittite Empire (Anatolian people) and the kingdom of Mitanni (Hurrian tribes) were destroyed. The Hittite kingdom was ruined by the so-called "people of the sea" invading from the Aegean Sea, as a result of which the Mushki, the ancestors of the Kartvelian tribes of the Meskhetians, settled east of the Hittites.

1.1 Ancient Georgian Kingdoms

At the end of the 3rd and at the beginning of the 2nd millennium BC in the South Caucasus were formed two political entities of Georgian tribes, which in scientific literature are known as kingdoms. Between the Likhi Ridge and the Black Sea was located the ancient Colchis Kingdom, whereas, in the east of the Likhi Ridge—Kingdom of Kartli (known as Iberia in the Greek-Roman literature) (Fig. 1.1). Later, the consolidation of these two Georgian kingdoms created the basis for the establishment of a Georgian country (Metreveli 1979).

At the same period by the end of the third and beginning of the second millennium BC, on the territory of Iberia a so called Trialeti culture began to develop, which is seen as a successor of the Kura–Araxes culture. The Trialeti culture has produced numerous gold artifacts of the highest artistic value. The Golden lion of Tsnori (Fig. 1.2) is among the examples thereof.

The archaeological collection of Vani encompasses a period between eighth and first centuries BC. It should be noted that the large number of unique gold artifacts unearthed in Vani (Fig. 1.3). That is why many people are apt to think that Vani is somehow linked to Argonauts journey and some regard it as a Golden Fleece city (Lordkipanidze 2001). However, factual data suggests that the Argonauts represented the Minoan civilization and should have therefore travelled to Colchis much earlier (by about 1000 years), Than Vani has reached the apex of its development during the Hellenistic period (Okrostsvaridze 2017).

1.1.1 Ancient Colchis

Internationally, Ancient Colchis is perhaps best known as the home of Medea and Golden Fleece. Geographically it covered the present-day Georgian provinces of Abkhazeti, Svaneti, Samegrelo, Imereti, Guria, Racha; Sochi and Tuapse districts of the modern-day Russia and Trabzon and Artvin provinces of present-day Turkey (Braund 1994).

Colchis was inhabited by a number of related but distinct tribes whose settlements lay along the shore of the Black Sea. The Kartvelian tribes differed so completely in language and appearance from the surrounding Indo-European nations that the ancients provided various "wild" theories to account for the phenomenon (Urushadze 1984).

According to the Greek sources, the ancient Colchis was a powerful kingdom. At the turn of the fifteenth and sixteenth centuries BC, the capital city of Colchis was Aea (the modern-day city of Kutaisi), where mighty Aeëtes reigned. Colchis has sort of "disappeared" further down the history, reappearing later, in the eighth century BC, when the kingdom regains strength after victorious wars with Sarduri II, king of Urartu in the years 750–742 BC. The following centuries saw the decline of ancient Colchis due to incessant invasions of Scythians and Cimmerians.

I. Gamkrelidze et al., *Geotourism Potential of Georgia, the Caucasus*, Geoheritage, Geoparks and Geotourism, https://doi.org/10.1007/978-3-030-62966-3_1

Fig. 1.1 Political map of the
Eastern Black Sea region in sixth
—second centuries BC.
Copyright © 2012 Deu

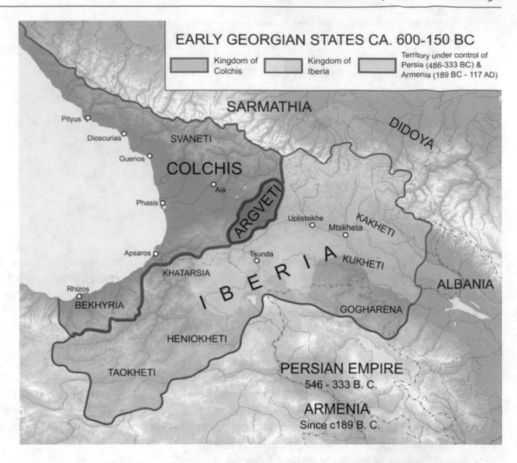

From the end of the sixth century BC, the Black Sea coast comes into the spotlight of Greeks. Greeks have set up commercial ports of Gonio-Asparos, Phasis, Dioscurias (Sebastopolis) and Pityus along the Colchian coast of the Black sea. After this came the period of Greek colonization of Colchis (Metreveli 1979).

Fig. 1.2 Golden Lion figurine (L—4.1 cm; H—2.1 cm; W—2.5 cm). 2300–2000BC. Tsnori, Alazani Valley, East Georgia/Iberia. The Georgian National Museum's collection

Fig. 1.3 Clip Headdress decoration (H—6.5 cm; H—6.5 cm), fourth century BC from Vani, Ancient Colchis/Western Georgia. The Georgian National Museum's collection

After the heavy defeat of Darius III inflicted by Alexander the Great at the Battle of Gaugamela in 331 BC, the Persian Empire lost its influence on the kingdoms of Colchis and Iberia, giving them the opportunity to develop independently.

1.1.2 Formation of the Kingdom of Kartli

The formation of the Kingdom of Kartli is connected with the Hellenistic era. According to the legend preserved in "Kartlis Tskhovreba", the invaders appointed one of their commanders, Azo, as the ruler of Kartli. Azo turned out to be a strict ruler, he often harassed the population. By his order, the fortresses here were destroyed. Azo was opposed by Pharnavaz, the nephew of the ruler of Mtskheta, who was aided by Kuji Eristavi of Colchis and the army of the North Caucasians. After the victory, Pharnavaz declared himself king of Kartli.

To ensure the rule of the country Pharnavaz divided the kingdom into 8 saeristavo (province) and 1 saspaspeto. Saeristavo was territorial-administrative unit headed by Eristavi the subordinate of the king, while saspaspeto was a district ruled by a Spaspet.

Pharnavaz built Mtskheta and other cities destroyed by the Greek-Macedonian invasion. He also created the main deity—the cult of Armazi, the idol of Armazi was erected on the so-called Kartli mountain opposite Mtskheta, where a fortress was built and it was called Armazi fortress. The establishment of Armazi as a common deity was of great importance for the newly formed Georgian state.

Leonti Mroveli attributes introduction of Georgian as a single state language in Kartli and the creation of Georgian writing and literacy to Pharnavaz.

Apart from historical Kartli, the kingdom included a large part of Kakheti, Samtskhe, Javakheti, Kola, Artaani, Klarjeti, as well as a significant part of western Georgia (Adjara, Argveti, Egrisi, Svaneti).

The most important crossings over the Caucasus ridge were subject to the Kingdom of Kartli. The neighbors of Kartli were also interested in strengthening these crossings, as there was a constant threat of invasion by nomadic-warrior tribes from there. That political factor was well used by the rulers of Kartli to their advantage.

1.2 Unification of Georgia and the Golden Age

After the establishment of the separate kingdoms of Georgia, the struggle for the unification of the country begins. For a while, the archbishopric of Kakheti became stronger, then after the weakening of Kakheti, the Abkhazian kingdom became more powerful, but neither Kakheti nor Abkhazia succeed in its ambition to unite the state. Ioane Marushisdze was the Eristavi of the King of Abkhazia. He knew well that it was David III Bagrationi, who could unite Georgia. David Bagrationi, who held the Byzantine title of Kouropalates, was the ruler of Tao at that time.

In 975, David III Kouropalates gathered an army and occupied Shida Kartli, repulsing the troops of the Kakhetian kingdom. David gathered the nobility of Kartli in Uplishikhe and told them that Bagrat III was his foster-son and all should obey him. David left Bagrat as the ruler of Kartli and appointed his father Gurgen as his guardian.

During the same period, the situation in western Georgia became tense. The rule of the blind Theodosius brought anxiety and disorder to Abkhazia. In such a situation, again at the initiative of Ioane Marushisdze, it was decided to make Bagrat III King of Abkhazia. In 978, Bagrat was consecrated king of the Abkhazians in Kutaisi. King Theodosius was sent to Tao. Bagrat received the title of King of the Abkhazians.

In 1008, Bagrat's father Gurgen died and all his domains were also inherited by Bagrat. After that, Bagrat managed to annex Kakhet-Hereti in 1010. Bagrat III was already called "King of the Abkhazians and Iberians, of Tao, and of the Arranians and Kakhetians, and the great Kouropalates of all the East". In 1014, the first king of united Georgia, Bagrat III, died. The king is buried in the temple of Bedia built by him (Metreveli 1979).

1.2.1 David IV the Builder

David IV son of George II was a king of Georgia in 1089–1125. He was a great statesman and powerful military leader, who has a special place in the history of Georgia (Metreveli 1979).

David inherited a very heavy legacy: the country was ravaged by enemy invasions, the Seljuk Turks nomadic tribes settled in the conquered territories and the Georgian people were in danger of physical annihilation, the noble feudal often disobeyed the king.

In 1121, the Seljuk Sultan Mahmud II declared jihad to Georgia and sent a strong army led by one of his famous generals Ilghaz to defeat Georgia. Despite the significant numerical superiority of the Turks, the Georgians, led by David IV, managed to defeat the invaders at the Battle of Didgori. In 1122 David IV took over Tbilisi and restored it as the capital of Georgia.

In 1124 David finally conquered Shirvan, later that year he fought Armenian city of Ani off the Muslim emirs and became the King of Armenians, incorporating Northern Armenia into the lands of the Georgian crown, after which David IV gained the title the "King of Kings, of the Abkhazians, Iberians, Armenians, Arranians, Kakhetians, sword of the Messiah, emperor (basileus) of all the East" (Metreveli 1979).

King David IV died in 1125. Throughout the 36 years of his reign he managed to regained almost all Georgian lands expelling the Seljuk invaders and leaving the country as a dominant force in Transcaucasia. Transforming a scattered, demoralized country into a strong and unified state king David handed over to his heirs a country that stretched from `Nikopsia to Daruband'. After his death King David was called Agmashenebeli the builder for english. For his personal dignity and great merit before the country and the nation, the Georgian Orthodox Church canonized David the Builder as a saint (Fig. 1.4) and set February 8 as the day of his commemoration.

1.2.2 Queen Tamar

Tamar the daughter of George III, a representative of the Bagrationi royal dynasty, became king of Georgia in 1184. Tamar reigned in the period of Golden Age of Georgia and turned out to be a highly successful ruler (Fig. 1.5). Tamar held the title: "King of Kings and Queen of Queens of the Abkhazians, Iberians, Arranians, Kakhetians and Armenians, Shirvanshah and Shahanshah; Autocrat of all the East and the West, Glory of the World and Faith; Champion of the Messiah". (Metreveli 1979).

In 1179, King George III proclaimed his daughter Tamar as co-regent. In 1184, after his death, a difficult situation arose in the country; the feudal aristocracy began a struggle to regain lost political privileges. In 1185, a group of influential feudales (nobility's) against her will tense Queen Tamar to marry Yuri, the son of Andrei Bogolyubsky,

Fig. 1.5 Queen Tamar and her father King George III of Georgia on the Betania monastery twelfth century fresco (photography by Giorgi Barateli)

known as "George the Russian". Very soon in about two years Tamar was divorced from Yuri and he was expelled from Georgia. After that Queen Tamar married a young Ossetian prince, David Soslan for love, in about 1189. Despite alliance with Ossetia brought Georgia little political advantage their union became fruitful. They had 2 children and Soslan showed good military talent.

Fig. 1.4 Fresco of King David IV Builder. Gelati Monastery, twelfth century (photography by Vladimer Shioshvili)

Fig. 1.6 Map of the Georgian Kingdom of the thirteenth century. Copyright © 2014 Maproom

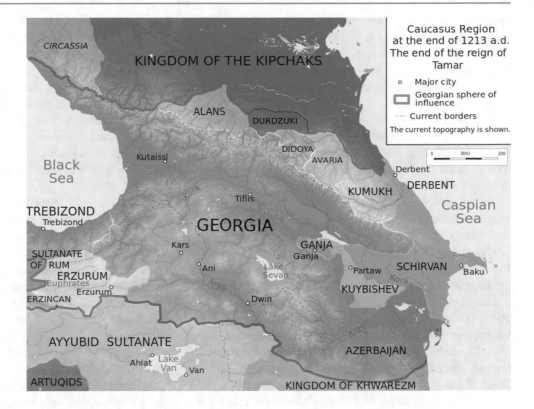

During Tamar's reign, Georgia became the strongest state in the Caucasus (Fig. 1.6). According to some historians, this strength was not based on the internal economic development of the country. They believe that the main source of the monarchy's economic strength was military booty and tribute, and that the unification of the country was unstable (Natadzc 1925; Rachvelishvili, 1927). According to other historians, the strength of a united Georgia was based on a strong socio-economic base base (Janashia 1943; Berdzenishvili 1965). Several important battles took place during the reign of Tamar, the most notable of which are the battles of Shamkori in 1195 and Basiani in 1203 (Java-khishhvili 1983). The victory of the Georgians in the Battle of Basiani was of great political significance; Georgia further strengthened its position in the region and completely strengthened its southern borders.

In twelfth and thirteenth centuries the education and culture of a united and economically strong feudal Georgia reached a high level. Georgian culture made a significant contribution to the political unification of feudal Georgia. From the beginning of the 12th century, secular writing was widely developed in Georgia, secular poetry was created. The Georgian national epic, "Vepkhistkaosani" (The Knight in the Tiger's Skin) written by Shota Rustaveli is especially noteworthy among the secular writing created during Tamar's period.

1.3 Georgia Under the Russian Empire

The "Friendship Agreement" between Russia and the United Kingdom of Kartl-Kakheti was signed on July 24, 1783 at the Georgievsk Fortress (North Caucasus), and as a result Kingdom of Kartl-Kakheti came under Russian protection. However, after the death of George XII in 1800, they violated the terms of the treaty and, with the manifestos of Emperors Paul I in 1800 and Alexander I in 1801 announced the abolition of the Kingdom of Kartl-Kakheti and the unification of Eastern Georgia with the Russian Empire. In 1804, the Russians invaded Imereti. In 1810 after several long battles, King Solomon II of Imereti was finally defeated. The Russians occupied Imereti kingdom and established Russian rule over it. Thus, the annexation of entire Georgia by Russian Empire was completed. After the annexation of Georgia, the Russians made every effort to uproot all remnants of Georgia's independence and to achieve the Russification of the Georgian people. To this end, they firmly followed of colonial policy upon their conquest, but they encountered the strong will of the Georgians to preserve their traditions, language, and culture. By the end of the nineteenth century this struggle had acquired a new, revolutionary character. The development of socialism began and socialist doctrines quickly spread in

Georgia. The National Movement merged with the socialist ideas and thus gave rise to the liberation movement, which became apparent after Russia, weakened by the First World War, gradually lost its influence over the territories it conquered. Their release became possible in 1917 as a result of the Russian Revolution.

The Russian Revolution began in Petrograd in March 1917. The king was overthrown and an interim government was formed to hold free elections of parliament, which in its turn decided on the status of Russia and all non-Russian countries within the empire.

Georgians and other nationalities living in Russian empire welcomed the overthrow of Tsarism and, most importantly, they remained loyal to the new democratic government. The Georgians did not try to secede from Russian empire; they were satisfied with the restoration of the independence of the Georgian Church and the opening of a Georgian university in Tbilisi. However, in November 1917, power in Russia was taken over by the Bolsheviks, which changed the situation dramatically. Georgians and other Caucasian peoples refused to recognize the Bolshevik regime, and on November 22, 1917, the Georgian National Council was established, with Noe Zhordania as a Chairman.

On March 3, 1918, the Bolsheviks signed the piece Treaty of Brest-Litovsk with the Union of Four, one of the provisions of which provided for the transfer of two Georgian (Batumi and Ardaghani) and one Armenian (Kars) districts to Turkey. The peoples of the Caucasus, of course, did not recognize this agreement and on April 9, 1918, declared the secession of Transcaucasia from Russia and the creation of an independent Transcaucasian republic consisting of Georgia, Armenia and Azerbaijan. It was the first voluntary federation in the history of the three Caucasian peoples. But on May 26, 1918, the Transcaucasian Republic disintegrated. On the same day, the Georgian National Council, which included representatives of all political parties and organizations, declared the independence of Georgia and the establishment of the Democratic Republic of Georgia. Thus, after 116 years, Georgia regained its rightful place in the family of free peoples.

1.3.1 Democratic Republic of Georgia

The National Council elected by the Georgian National Assembly on November 22, 1917, adopted the Act of Independence on May 26, 1918 in Tbilisi, which was officially ratified on March 12, 1919 by the Constituent Assembly.

The proclamation of the First Republic of Georgia took place against the backdrop of strong pressure from the Ottomans, whose army had invaded the heart of Georgia, a few kilometers from Tbilisi. It was possible to stop the

Ottoman intervention after the German demarche, which had its own interests in Georgia. On June 4, 1918, the Treaty of Batumi was signed, which established the borders between the Ottomans and Georgia: the Ottomans were given the districts of Batumi and Kars, the provinces of Akhalkalaki and Akhaltsikhe. At the end of June, a German military expedition arrived in Poti. The Germans received unlimited access to local resources from the Georgian government. From June to September 1918, they exported 30 million marks worth of manganese, copper, tobacco, bread, tea, fruit, wine and more.

Throughout its four-year history, the Democratic Republic of Georgia has always been involved in hostilities as part of the border demarcation process. However, armed confrontations were taking place not only with neighboring countries, but also within the republic. The aim of the Bolsheviks in Abkhazia, Samegrelo and South Ossetia was to overthrow the Georgian Democratic Republic and unite it with Soviet Russia, but the Georgian army managed to neutralize them. In 1920, the Bolsheviks tried to stage a coup in Tbilisi, however to no avail. At the same time, Georgian troops defeated the Soviet army encamped on the border with Azerbaijan, thereafter which Soviet Russia recognized Georgia's independence, as reflected in the May 7 Moscow Agreement.

The new government of Georgia experienced revolution, counter-revolution, civil war, anarchy, the fall of the national currency and the economic crisis. After the establishment of order and return to normal life, the National Council set out to prepare for the election of the Constituent Assembly. Elections were held in February 1919 on the basis of a direct, equal, universal, secret and proportional electoral system. The main task of the Constituent Assembly was to draft the Constitution of the Republic. The task of entrusting this task to the Emergency Committee.

The Constitution of Georgia was finally adopted on February 21, 1921. At that time, fierce fighting was taking place on the outskirts of Tbilisi between the Georgians and the Soviet Army. On February 25, the Red Army captured Tbilisi and announced the creation of the Soviet Socialist Republic of Georgia. The government of the Democratic Republic of Georgia moved to Batumi leading the Georgian army from there. In March, the Turkish army invaded Adjara. The Georgian government left the country and emigrated.

1.3.2 Soviet Socialist Republic of Georgia

At the end of March 1921, the Red Army completely controlled the territory of Georgia. In 1922, Georgia became a member of the Soviet Union. The 1920s and 1930s were a period of establishment of the in Georgia. Joseph Stalin

Fig. 1.7 The flag of Georgia and the coat of arms of Georgia

(Jugashvili), an ethnic Georgian, was born in Gori. He was one of the prominent figures among the Bolsheviks. Stalin was leading the Soviet Union from the mid–1920s until his death on 5 March 1953. The Soviet leadership carried out a series of economic reforms aimed at the rapid industrialization of the semi-agrarian country. Industrialization in Georgia has not been very successful. The level of industry by 1925 could not be equal to the level of 1914. At the same time, agriculture was being collectivized, implying the establishment of peasant associations characteristic for the socialist set-up. Collectivization led to great resistance from rich and affluent peasants (so-called "kulaks"). "Gankulakeba" (dispossession of kulaks) went very difficult in Georgia.

World War II broke out in 1939, in which the Soviet Union suffered the greatest material and human losses. 700 thousand men were called to war from Georgia and 300 thousand died among them. In this brutal war, one dictator—Joseph Stalin defeated another dictator—Adolf Hitler.

From 1953 to 1964, Nikita Khrushchev carried out several reforms. Market relations were being established in the economy of the Soviet Union. Pseudo-capitalist shadow economy coexisted with the socialist economy in Georgia. The level of corruption was high.

In 1985, a significant change took place in the Soviet Union. Mikhail Gorbachev became the ruler of the USSR. Gorbachev acknowledged the country's deep crisis and set out to rectify it, but it was impossible. National movement became widespread in Georgia as well as in other Soviet republics.

1.4 Independent Georgia

On April 9, 1989, Soviet troops dispersed peaceful protesters in Tbilisi. On March 31, 1991, a referendum on Georgia's independence was held. 98% of the referendum participants supported the independence of Georgia. Zviad Gamsakhurdia became the first President of independent Georgia.

Modern Georgia is a country in the Caucasus region of Eurasia. Located at the crossroads of Western Asia and Eastern Europe. It covers a territory of 69700 km^2, with 2017 population of 3.718 million in 2017. "Georgia is a parliamentary representative democratic republic with a multi-party system. The President of Georgia is the ceremonial head of state and the Prime Minister of Georgia is the head of government" (Georgian Parliament 2018). The capital and the largest city of Georgia is Tbilisi. The flag and coat of arms of Georgia (Fig. 1.7) are the national symbols of the country. The modern flag comes from the banner of the medieval Georgian kingdom. The idea of the coat of arms was based on the Bagration family coat of arms, but it was modernized and adopted a relatively simple composition, currently featuring St. George, the traditional patron saint of Georgia, on a purple shield.

In August 2008, because of tensions between Russia and Georgia, Russia began the war against the Georgia. Since the war, Abkhazia and South Ossetia are the Georgian territories occupied by Russia.

1.5 Cultural Heritage of Georgia

Georgia is known for its original script, traditional dance, polyphonic folk music and Georgian original cuisine.

1.5.1 Georgian Scripts

In the opinion of Vakhtang Licheli, a Georgian archaeologist: "Mysterious script found on the Grakliani Hill (Fig. 1.8) may turn out to be the oldest example of native writing found in the Caucasus—entire thousand years older than any indigenous writing previously found in the region" (Licheli 2010).

Georgian script—an alphabetic script used by the Georgian language and its related Kartvelian languages (Megrelian, Lazian and Svanian). The modern Georgian alphabet

Fig. 1.8 Mysterious script of the Grakliani Hill (photography by Shalva Lezhava)

ა	ბ	გ	დ	ე	ვ	ზ	თ	ი	კ	ლ
ani	*bani*	*gani*	*doni*	*eni*	*vini*	*zeni*	*tani*	*ini*	*k'ani*	*lasi*
მ	ნ	ო	პ	ჟ	რ	ს	ტ	უ	ფ	ქ
mani	*nari*	*oni*	*p'ari*	*zhani*	*rae*	*sani*	*t'ari*	*uni*	*pari*	*kani*
ღ	ყ	შ	ჩ	ც	ძ	წ	ჭ	ხ	ჯ	ჰ
ghani	*q'ari*	*shini*	*chini*	*tsani*	*dzili*	*ts'ili*	*ch'ari*	*khani*	*jani*	*hae*

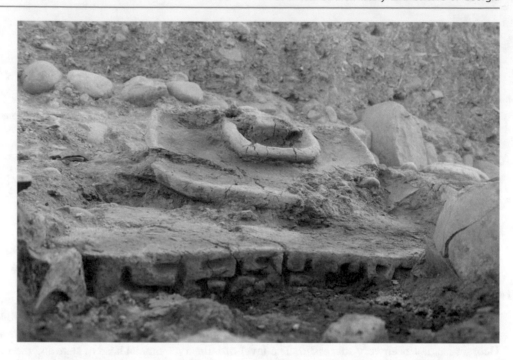

Fig. 1.9 The Georgian scripts of Mkhedruli. Copyright © 2016 Nick

has 33 letters, while the old alphabet had 38 letters, five of which are no longer used in modern Georgian.

According to Kartlis Tskhovreba (literally "Life of Kartli"), the Georgian script was created by the first king of Iberia, Pharnavaz.

At the initiative of the Georgian National Agency for Cultural Heritage Preservation, three types of the Georgian alphabet—Asomtavruli, Nuskhuri and Mkhedruli (Fig. 1.9)—were awarded the status of Intangible National Cultural Heritage Monument. In March 2015 it was inscribed on the UNESCO World Heritage List.

1.5.2 Traditional Georgian Dance

Georgian folk choreography was formed at the crossroads of two cultural worlds - Europe and Asia. It has a history of many centuries. The survived archaeological and ancient literary monuments confirm that the historical precursors of Georgian folk choreography were cult dances. One of the earliest patterns of cult dances was a hunting dance, performed in honor of the god of fertility - the moon ("Shushpa").

A figure of a dancing woman (sixth century BC) represented on a bone plate discovered during the excavation of Bagineti indicates that ritual dances of women were also held near the temple of the god of fertility.

The development of agriculture and animal husbandry was followed by the establishment of new customs, which were reflected in the ritual dances of a magical nature.

There also has long been a partner dances associated with the cult of fertility, which were an accompanying element of the great festivals of pagan times.

The combination of purely dance elements with customs and traditions and game rituals led to the creation of dances such as: "Kartuli", "Gandagana", "Khorumi", "Samaia",

"Khanjluri", "Mtiuluri", "Davluri", "Bagdaduri" and others. These dances were a kind of "discovery" in Georgian choreography. Over time, many dance troupes were formed, which widely popularized Georgian choreography inside the country as well as abroad.

1.5.3 Georgian Folk Music

Georgian folk music dates back to ancient times. It was from the beginning connected with the existence of our ancestors. It reflects like a mirror the richest history of the Georgian nation, spiritual aspirations, moral strength, patriotic and heroic attitude. Folk music has always been the spiritual nourishment of the nation and has been associated with everyday activity of a Georgian man. People sang in times of war, labor, sickness, joy, sorrow. Singing was a force that added courage and resilience in adversity. That is why the Georgian man was careful with his centuries-old musical language.

Ancient information about Georgian folk music can be found in archeological materials and ancient Greek sources. Various Georgian instruments founded during archaeological excavations - salamuri (flute), changi (like harp), lyre are supposedly dated back to the fifteenth - thirteenth centuries BC.

It should be noted that Georgian folk songs have been characterized by polyphony since time immemorial. Polyphony is considered to be a unique phenomenon in the world. Georgian polyphony among them is completely special, different and multifaceted. Every corner of Georgia has its own musical dialect, there are up to 15 such dialects. UNESCO as an Intangible Heritage masterpiece recognized Georgian vocal polyphony in 2001.

Georgian folk songs have been refined and enriched for centuries. Georgia is the homeland of "Chakrulo", "Lile", "Khasanbegura", "Mravaljamieri", "Iavnana" and many other brilliant songs. "Chakrulo" was one of 29 musical compositions included on the Voyager Golden Records that were sent into space on Voyager 2 on 20 August 1977.

1.5.4 Georgian Cuisine

Georgian cuisine is a culinary style that originated in Georgia and has been established in the country for centuries. There is a noticeable difference between the regional cuisines of the country. There is an abundance of meat, wine, brad and animal fats in Eastern Georgia, while the cuisine of Western Georgia is characterized by sauces mixed with walnuts and peppers, dishes of meat and dairy products.

Georgia is the homeland of wine, vines grow almost everywhere here, Georgian wine is put in pitchers.

The world's oldest wines, are made in ancient authentic Georgian underground clay wessels—"Qvevri". The fertile valleys and protective slopes of the Transcaucasus were home to grapevine cultivation and Neolithic wine production for at least 8000 years (McGovern et al. 2017).

Therefore, Georgia bears the status of "the Cradle of Wine". More than 500 grape varieties grow here and contemporary winemaking develops intensively to satisfy international market. In 2013, UNESCO added the ancient traditional Georgian winemaking method using the clay wessels to the UNESCO intangible Cultural Heritage Lists (UNESCO Report 2017).

1.6 Summary

Georgia is a country with ancient history and culture. At the end of the 3rd and at the beginning of the 2nd millennium BC in the South Caucasus were formed two political entities of Georgian tribes: ancient Colchis Kingdom in the west and Kingdom of Kartli (Iberia) in the East creating the basis for the establishment of a Georgian country. During the eleventh to twelfth centuries under King David IV the Builder and Queen Tamar the Great Georgia passed the so-called "Golden Age" of its history. Later the country underwent the invasions of the Mongols (at the beginning of 1220 s) and Persians (in the fifteenth century). After that, Georgia was conquered by Russia for a long time and only for a very short independent Democratic Republic. In 1921–1990 it was Georgian Soviet Socialist Republic as part of the USSR. Since 1991 Georgia is independent country. In August 2008, Russia began the war against the Georgia. Since the war, Abkhazia and South Ossetia are Russian-occupied territories of Georgia.

Georgia is known for its unique scripts, traditional dance, polyphonic folk music and Georgian original cuisine.

References

Berdzenishvili N (1965) Issues in Georgian History 2, Tbilisi. (in Georgian)

Braund D (1994) Georgia in antiquity: a history of Colchis and Transcaucasian Iberia, 550 BC-AD562. Oxford, USA. Oxford University Press, pp 1–378

Gabunia L, Vekua A, Lordkipanidze D, Swisher et al (2000) Earliest Pleistocene Hominid Cranial Remains from Dmanisi, Republic of Georgia: taxonomy, geological setting, and age. Science 288 (5468):1019–1025. https://doi.org/10.1126/science.288.5468.1019

Georgian Parliament (2018) Constitution of Georgia; http://www.parliament.ge/en/kanonmdebloba/constitution-of-georgia-68

Janashia S (1943) History of Georgia from ancient times to the end of 19-th Century. Tbilisi. (in Georgian)

Javakhishvili I (1983) History of Georgian Nation, Tbilisi. (in Georgian)

Licheli V (2010) Grakliani Hill. In: Cultural and trade relations in the Central Transcaucasus. Tbilisi, pp 25–88

Lordkipanidze O (2001) "The Golden Fleece": myth, euhemeristic exploration and archeology. Oxford J Archeol 20:1–38

McGovern P, Jalabadze M, Batiuk S, Callahan MP, Smith KE, Hall GR, Kvavadze E, Maghradze D, Rusishvili N, Bouby L, Failla O, Cola G, Mariani L, Boaretto E, Bacilieri R, This P, Wales N, Lordkipanidze D (2017) Early Neolithic wine of Georgia. Proc Natl Acad Sci 114(48):E10309–E10318. https://doi.org/10.1073/pnas.1714728114

Metreveli RV (1979) The History of Georgia: a dictionary. Publishing House "Ganatleba", Tbilisi, pp 1–351 (in Georgian)

Natadze G (1925) Brief sociological analysis of Georgian History. Tbilisi. (in Georgian) http://217.147.235.82/bitstream/1234/314594/1/SakartvelosIstoriisMokleSociologiuriMimoxilva.pdf

Okrostsvaridze A (2017) The Argonauts: a modern investigation of the mythical "Gold Sands". LAMBERT Academic Publishing, Saarbrücken, p 92

Rachvelishvili K (1927) Feudalism history of Georgia. Tbilisi. (in Georgian) http://www.dspace.gela.org.ge/bitstream/123456789/6851/1/Sakartvelos%20peodalizmis%20Istoria-1926.pdf

UNESCO Analytical and Technical Report (2017) UNESCO Culture for development indicators

Urushadze A (1984) The Country of the enchantress Medea. Publishing House "Mecniereba", Tbilisi, pp 1–129 (in Georgian)

2.1 Introduction

The main orographic units of the Caucasus are: The Greater Caucasus mountain chain, intermountain area divided by Likhi Ridge into the Colchis (Rioni) and Iberian (Kura) lowlands and the Lesser Caucasus mountain chain (Fig. 2.1).

The climate in Georgia is diverse due to its location in the subtropical zone at a boundary of Aral-Caspian arid region and continental highland Caucasian (Gamkrelidze 1997b). Western Georgia is characterized by a moist subtropical climate, the Eastern one by dry. The climate in South Georgia is mainly continental (Gamkrelidze 1997b).

Network of rivers in West Georgia, belonging to the Black Sea Basin, is denser than in its eastern part (Gamkrelidze 1997b). In Western Georgia main rivers are: the Rioni, Enguri, Kodori, Bzibi, Tskenistskali (the right tributary of the Rioni). The largest river of Georgia and Caucasus is the Mtkvari (Kura) with an outflow in Turkey. The main tributaries of one are the Liakhvi, Ksani, Aragvi, Khrami, Iori and Alazani. To the Caspian Sea Basin also belongs the River Tergi, which for 85 km flows from south to north and crosses the northern border of Georgia with Russia. The capital of Georgia is Tbilisi. Other significant settlements are: Kutaisi, Sokhumi, Poti, Gagra, Borjomi, Gori, Mestia, Kazbegi, Telavi and Lagodekhi (Fig. 2.1).

In a geological sense, Georgia is a component of the Caucasian segment of the Mediterranean (Alpine-Himalayan) collisional orogenic belt (Gamkrelidze et al. 2015). It was formed after opening and closing processes of Proto-, Paleo- and Neo-Tethys oceans, within which terranes (relatively small continental lithospheric plates) of different age and geological history were situated. The latter eventually joined the Eurasian continent forming the present Mediterranean orogenic belt (Gamkrelidze 1997a, 2016) (see below Sect. 2.5).

Currently, the Caucasus is in a state of a continental collision between the Arabian plate and Scythian (East European) platform, which leads to current various geological processes in the region: vertical and horizontal movements of separate blocks, earthquakes, volcanic activity and etc.

The geological structure of the territory of Georgia has been studied for more than 200 years. In the beginning, it was research mainly by foreign geologists. At the beginning of the twentieth century, especially after the founding of the geological institute in Tbilisi in 1925 by the creator of the Georgian Geological School academician Aleksandre Janelidze, research was carried out mainly by Georgian geologists. The result of these works was a fundamental monograph on the geological structure of Georgia (Geologia SSSR, v. X, Gruzinskaia SSR, 1964). This book examined issues of stratigraphy, intrusive and effusive formations, tectonic structure, seismotectonics and the history of the geological evolution of the territory of Georgia. The authors of the book were prominent representatives of the first generation of Georgian gelogists: Peter Gamkrelidze (editor), Ivane Kacharava, Giorgi Dzotsenidze, Archil Tsagareli, Mikheil Eristavi, Giorgi Zaridze, Nikoloz Skhirtladze and others.

After that, the relatively young generation of Georgian scientists worthily continued the study of all aspects of the geology of Georgia and the entire Caucasus mainly in the light of the Plate Tectonics. The results of these studies are published in many scientific papers, some of which are cited in this book.

The territory of Georgia covers all major tectonic units (zones) of the Caucasian orogeny: The Greater and the Lesser Caucasus fold systems and intermountain depressions lying between them. This area represents a real "natural geological laboratory" exposing magmatic, sedimentary and metamorphic rocks, ranging wide on the geologic time scale (from the Neoproterozoic to the Quaternary inclusive). From the Late Miocene (9–8 Ma), intensive exhumation of the Caucasus orogen began, during which molasse, lacustrine and subaerial volcanic formations were being accumulated (Gamkrelidze 1997b). These rocks keep of the history of geological processes that have built numerous geological sites of important geological information observed in the

I. Gamkrelidze et al., *Geotourism Potential of Georgia, the Caucasus*, Geoheritage, Geoparks and Geotourism, https://doi.org/10.1007/978-3-030-62966-3_2

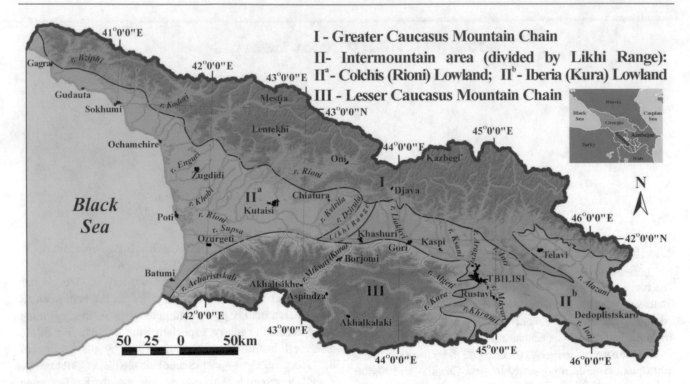

Fig. 2.1 Main orographic units of Georgia modified after Maruashvili (1970); Topographic map from the United States Geological Survey (USGS) earth explorer web-based platform https://earthexplorer.usgs.gov

present-day Georgia and are excellent destination for development of geotourism industry (Gamkrelidze et al. 2011a, 2012, 2019).

The diversity of geologic objects is not the only reason Georgia can be fascinating. It is also home to significant paleoarchaeological, archaeological and historic sites that can be visited concurrently during the geological tours (Gamkrelidze et al. 2019). Mtskheta, the capital of the early Georgian Kingdom of Iberia (during 3rd BC–4th AD), and Dmanisi paleoanthropological site (1.8 Ma) are worthy of special mention. In addition to this, it should be noted that all potential geopark sites and geotourist routes contain some of the important cultural heritage as well.

2.2 Stratigraphy and Rock Types of the Different Tectonic Zones of Georgia

2.2.1 Precambrian and Paleozoic Rocks

The oldest, Precambrian and Lower-Middle Paleozoic rocks are exposed in all the structural units of Georgia (Figs. 2.2 and 2.3). In particular they are represented within the Main Range zone of the Fold system of the Greater Caucasus, the Transcaucasian microplate (the Dzirula massif), and the fold system of the Lesser Caucasus (the Khrami and Loki

massifs) (Gamkrelidze and Shengelia 2005; Gamkrelidze et al. 2019).

To the Precambrian belong so-called Buulgen and Gondaray metamorphic complexes of the Main Range zone of the Greater Caucasus (Gamkrelidze et al. 2020). According to U-Pb LA-ICP-MS data they experienced Cadomian stage of high temperature regional metamorphism with weighted mean ages 626 ± 2 and 627 ± 19, Caledonian stage of high temperature prograde regional metamorphism with mean ages 461 ± 5.3 Ma, 457 ± 12 and Late Variscan regressive regional metamorphism with mean ages 312.5 ± 4 and 317.0 ± 8.3 Ma (Gamkrelidze et al. 2020).

In the Dzirula massif the Precambrian rocks are represented by gneiss-migmatite complex experienced Cadomian stage of regional metamorphism with U-Pb LA-ICP-MS zircon age 650–540 Ma (Gamkrelidze et al. 2011b). But the gneiss-migmatite complex bears signatures of the earliest (Grenvilian) metamorphism in the Dzirula massif, because the rocks of this complex are cut by Precambrian quartz-diorite gneiss (orthogneiss) that carry the xenoliths of the metamorphosed rocks of the complex, including strongly deformed plagiomigmatites and crystal schists (Gamkrelidze and Shengelia 2005, 2007; Gamkrelidze et al. 2011b) (Fig. 2.4).

Paleozoic rocks in the Main Range zone of the Greater Caucasus are represented by Laba metamorphic complex consisting of four allochthonous units: Mamkhurts, Damkhurts, Lashtrak, and Adjarka. The meta-volcanic to

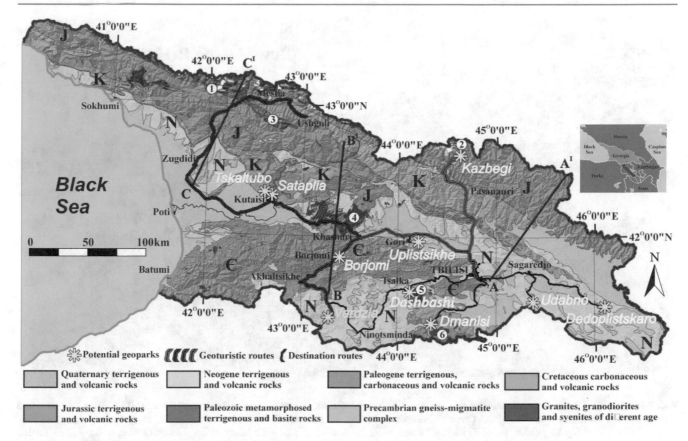

Fig. 2.2 Simplified Geological map of Georgia (adapted after Gudjabidze and Gamkrelidze 2003), with geotouristic routes and potential geopark locations. Numbers in the circles: exposures of pre-Alpine crystalline basement: 1—Greater Caucasus, 2—Dariali; 3—Paleozoic sedimentary weakly metamorphic rocks of Dizi series, 4—Dzirula, 5—Khrami, 6—Loki; A–A^I, B–B^I, C–C^I—Lines of geological cross-sections (the last ones see below)

meta-sedimentary Damkhurts unit contains carbonate layers with Siluro–Devonian fossils and polymict stretched-pebble conglomerates with diverse pebble compositions. The Adjarka unit contains quartzite, amphibolite, porphyry, and fossiliferous limestones with post-Ordovician crinoids (Gamkrelidze and Shengelia 2005, 2007; Gamkrelidze et al. 2019).

Paleozoic rocks are exposed also in the central part of the southern slope of the Greater Caucasus. They are represented mainly by black shales, sandstones, turbidites, olistostromes, lenses of marble and calc-alkaline andesite-dacitic volcanoclastics. Their visible thickness reaches 2000 m. This is the so-called Dizi series, in which faunally (by corals, foraminifera, and conodonts) the Devonian, Carboniferous, and Permian are established (Gamkrelidze and Shengelia 2005, 2007; Gamkrelidze et al. 2019) (see below Sect. 3.2, Stop 24).

Comparatively weakly metamorphosed Paleozoic sediments are exposed in the Dzirula massif as well. These are the allochthonous plates of the so-called "phyllitic suite", which are in contact with Upper-Paleozoic granitoid and Precambrian gabbro-amphibolite and serpentinite (the metaophiolites) (Gamkrelidze et al. 1981) (see below Sect. 3.2, Fig. 3.23). Precambrian and Paleozoic ophiolites within the crystalline core of the Greater Caucasus and in Lok-Karabakh zone (in the Loki massif) are present as well (Gamkrelidze and Shengelia 2005, 2007; Gamkrelidze et al. 2019).

The Upper Paleozoic rocks are also developed in all tectonic units of Georgia. In the Main Range zone of the Greater Caucasus, crystalline rocks are overlain by weakly metamorphosed sandstones, conglomerates, and argillites, which contain Upper Carboniferous-Lower Permian marine fauna.

Continental and coastal calc-alkaline rhyolitic volcanic rocks and coal-bearing argillites with lenses of reef limestones are known in the Dzirula and Khrami massifs. Lower-Middle Carboniferous corals, brachiopods, foraminifers, and terrestrial flora have been found in this formation in the Khrami massif (Gamkrelidze and Shengelia 2005, 2007).

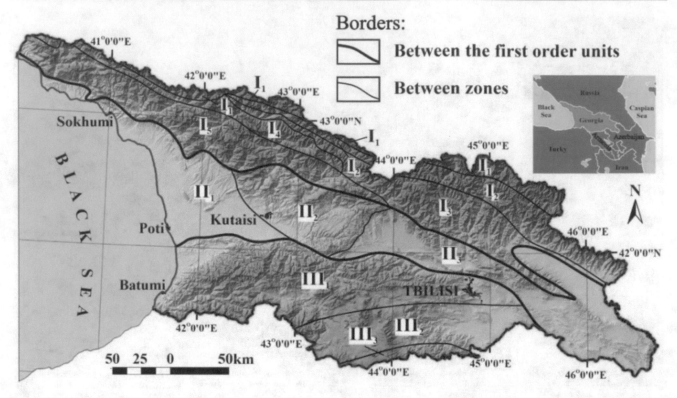

Fig. 2.3 Simplified tectonic subdivision of the territory of Georgia (Gamkrelidze 2000). I-Fold system of the Greater Caucasus: I_1-Main Range zone; I_2-Kazbeg-Lagodekhi zone; I_3-Mestia-Tianeti zone; I_4-Chkhalta-Laila zone; I_5-Gagra-java zone; I_6-Novorossiysk-Lazarevskoe zone. II-Transcaucasian intermountain area: II_1-Western molasse zone of sinking; II_2-Central zone of uplift; II_3-Eastern molasse zone of sinking. III-Fold system of the Lesser Caucasus: III_1-Adjara-Trialeti folded zone; III_2-Artvin-Bolnisi zone (block); III_3-Lok-Karabakh zone

Fig. 2.4 Xenolith of migmatized postcrystallization-folded biotite crystalline schist in Precambrian quartz-diorite gneiss (Dumala river gorge) (Gamkrelidze 1976)

2.2.2 Mesozoic and Cenozoic Formations

The biostratigraphy of Mesozoic and Cenozoic deposits of Georgia is very well studied. Fundamental in this area are the works of the following researchers, most of whom are not alive: A. Janelidze, I. Kakhadze, I. Kacharava, A. Tsagareli, M. Eristavi, K. Nutsubidze, N. Bendukidze, N. Khimshiashvili, R. Gambashidze, V. Zesashvili, T. Lominadze, M. Topchishvili, E. Kotetishvili, N. Tsirekidze, I. Kvantaliani, M. Kakabadze, M. Sharikadze, D. Buleishvili,

G. Chelidze, G. Ananiashvili, M. Uznadze, O. Janelidze, L. Maissuradze, I. Shatilova and others.

Mesozoic and Cenozoic formations are developed in all tectonic units of Georgia (Gamkrelidze 1997a, b, 2000, 2016; Caputo et al. 2000; Gudjabidze and Gamkrelidze 2003; Gamkrelidze et al. 2015, 2019).

Biostratigraphy and rock types of Mesozoic and Cenozoic deposits of the different tectonic zones of Georgia recently are described briefly by Gamkrelidze (2016) and Gamkrelidze et al. (2019).

Triassic and Lower Jurassic deposits are developed in all tectonic units of Georgia (Fig. 2.2) and are represented mainly by terrigenous strata, while the Middle Jurassic (Bajocian) is composed of thick submarine volcanic rocks. In particular, of calc-alkaline basalts, andesite-basaltes, andesite-dacites and their pyroclastics.

The Upper Jurassic on the southern slope of the Greater Caucasus is composed of flysch rocks, and on the Georgian block of lagoonal-continental terrigenous deposits.

The Lower Cretaceous in all tectonic zones is represented mainly by carbonate deposits (Fig. 2.2). The Upper Cretaceous consists mainly of submarine volcanic rocks with calc-alkaline basaltic composition, as well as of carbonate rocks.

Paleogene deposits are developed in almost all tectonic zones (Fig. 2.2) and are represented mainly by terrigenous rocks, as well as in the Middle Eocene by submarine volcanic rocks with calc-alkaline basalt-andesite-dacite-rhyolite composition.

Neogene deposits are developed mainly in the Transcaucasian intermountain area and are represented in the lower part (Middle Miocene—Middle Sarmat) by marine molasses. However, in the Artvino-Bolnisi and Lok-Karabakh zones, they are composed of subaerial calc-alkaline andesites, andesite-dacites and dolerites. Their upper part includes the Pleistocene and Quaternary too (Fig. 2.2).

Mesozoic and Cenozoic deposits are characterized here in detail by means of stratigraphic columns, showing the stratigraphic units of separate tectonic zones of Georgia, their thickness, interrelation and lithology (Figs. 2.5, 2.6 and 2.7).

2.3 Intrusive Rocks

Intrusive rocks in pre-Alpine crystalline basement of Georgia are represented mainly by Precambrian orthogneisses (in the Dzirula massif), supra-subduction pre-Variscan and Early Variscsn syn-metamorphic plagiogneisses gneissic quartz-diorites, plagiogranites and granodiorites and Late Variscan microcline granites, which include the following:

(1) Granitization products of the quartz-diorite gneisses and other ancient rocks; (2) Magmatic massive even-grained homogeneous (eutectic) granites, alaskites, and aplites; (3) Coarse-grained porphyritic granites of the so-called Rkvia intrusion exposed in the western part of the Dzirula crystalline massif (Gamkrelidze and Shengelia 2005, 2007).

Among the Mesozoic and Cenozoic intrusives one can distinguish supra-subduction, intraplate and collisional types of plutons (Gamkrelidze et al. 2002). Supra-subduction plutons are represented by Middle Jurassic intrusives of the Greater Caucasus, Dzirula and Loki Massifs, and Eocene intrusives of the Adjara-Trialeti zone.

Eocene intrusives of the Adjara-Trialeti zone are represented by subvolcanic bodies and small stokes of gabbro-essexites, gabbro-monzonites, monzonites, quartz-diorites, syenites and alkaline syenites (Gamkrelidze et al. 2002).

To the intraplate intrusives belongs teschenite complex of western Georgia, which is represented by teschenites and accompanying veined bodies of essexites, camptonites and monchikites (Skhirtladze 1958).

Collisional type of plutons is represented by Neogene diorites and Pliocene-Quaternary granit-porphires in Lower Svanetia (Tsana intrusion), in Upper Racha (Karobi intrusion) and in Khevsureti (Blo intrusion) (Gamkrelidze et al. 2002).

2.4 Tectonic Structure of Georgia

The territory of Georgia, as a part of the Caucasus, underwent a long and complicated tectonic evolution and contains structures of various types, scales and genesis (Gamkrelidze et al. 2013, 2019; Gamkrelidze 2016).

The pre-Alpine crystalline basement of Georgia is characterized by the development of deep seated tectonic nappes, which are widespread in the Main Range zone of the Greater Caucasus, in the Dzirula and Loki massifs (Gamkrelidze 1991, 2016; Gamkrelidze and Shengelia 2005).

Alpine structures have a different character in the various tectonic zones. The Foldsystem of the Greater Caucasus, is characterized by a distinctly expressed asymmetry in its structure: southward verging, often isoclinal folding on the southern slope and quiet, poorly folded, or monoclinal structure on the northern slope (Gamkrelidze 1991, 2016; Mauvilly et al. 2015, 2016). Large southward-directed nappes are developed also on its southern slope (Gamkrelidze and Gamkrelidze 1977; Gamkrelidze 1991, 1997b, 2016; Gamkrelidze et al. 2019) (Fig. 2.8). Within it the Alisisgori-Chinchvelta, Sadzeguri-Shakhvetila and Zhinvali-Pkhoveli nappes and the Ksani-Arkala parautochthone are distinguished, representing in geological past independent structural-facies zones.

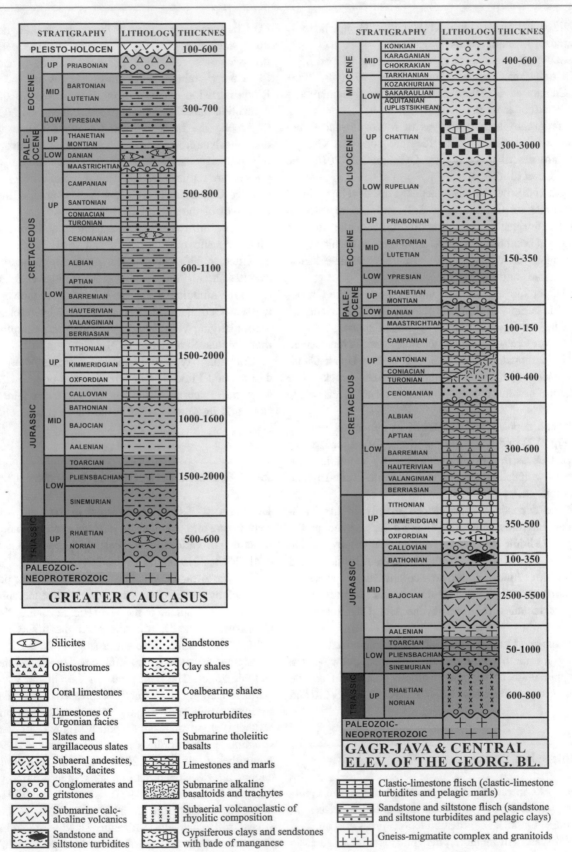

Fig. 2.5 Stratigraphic columns of the Fold system of the Greater Caucasus, Gagra-Java zone and Central elevation of the Georgian block

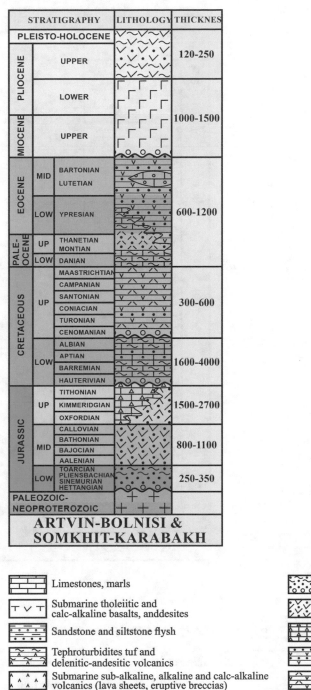

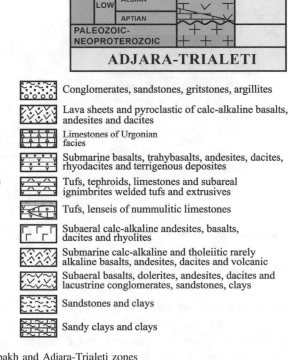

Fig. 2.6 Stratigraphic columns of the Artvin-Bolnisi, Somkhit-Karabakh and Adjara-Trialeti zones

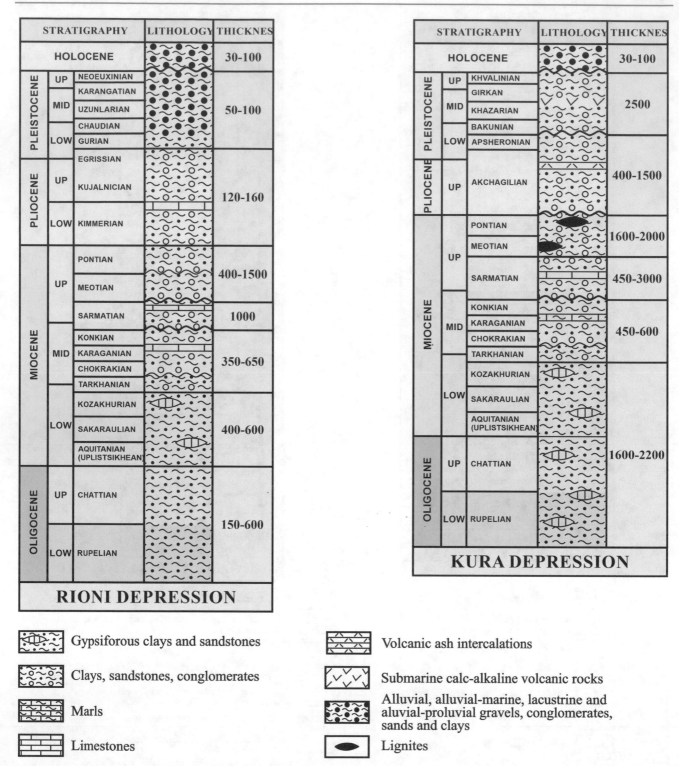

Fig. 2.7 Stratigraphic columns of the Rioni and Kura depressions

The northern boundary of the Georgian Block, in its western part, is formed by a deep fault, which in the sedimentary cover manifests itself as a regional flexure. Study of the structural peculiarities of the Georgian Block has shown that its central and western parts are characterized by a mosaic-block structure of the basement and occurrence of typical above-fault folds in the sedimentary cover. In the eastern part of subsidence of the Georgian Block, its cover is detached and shifted towards the south together with the nappes of the southern slope of the Greater Caucasus (Gamkrelidze 1991) (Figs. 2.8, 2.9 and 2.10).

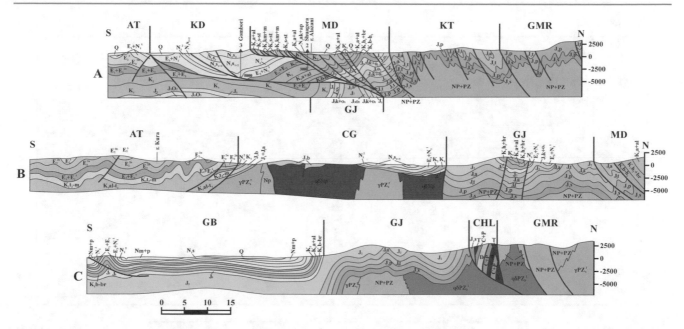

Fig. 2.8 Geological cross-sections of the territory of Georgia: AA[1]—in the eastern part, BB[1]—in the central part, CC[1]—in the western part (positions of sections see in Fig. 2.2; Stratigraphyc signes see below in Fig. 3.10). *Tectonic units* GB—Georgian block, GJ—Gagra-Java zone, CHL—Chkalta-Laila zone, GMR—Greater Caucasus Main Range zone, AT—Adjara-Trialeti zone, CG—Central elevation of the Georgian block, MT—Mestia-Tianeti zone, KL—Kazbeg-Lagodekhi zone

The Adjara-Trialeti zone of the Lesser Caucasus is, on the whole, an anticlinorium and is characterized by block-fold and thrust-fold structures. To the west from the Dzirula massif along the northern margin of this zone, an overthrust nappe is developed (Basheleishvili et al. 1982) (Fig. 2.8).

The Artvini-Bolnisi zone consists of two different tectonic units: the Javakheti zone (in the west) and the Bolnisi zone (in the east) (Fig. 2.3). In the young (Neogene-Pleistocene) volcanic cover of the Javakheti zone, sub-latitudinal gentle folds are observed.

The Bolnisi zone includes the horst-like Khrami inlier of pre-Alpine basement and the territory covered with Cretaceous and Paleogene volcanic rocks. Brachyanticlines and steep faults of various orientations are developed to the south in a sedimentary cover, which generally forms a gentle syncline (Gamkrelidze 1997b, 2000, 2016; Gudjabidze and Gamkrelidze 2003).

The northeastern wedge of the Lok-Karabakh zone forms part of Georgia (Fig. 2.3) and is characterized by echelon-like disposition of internal anticlinoria. In the core

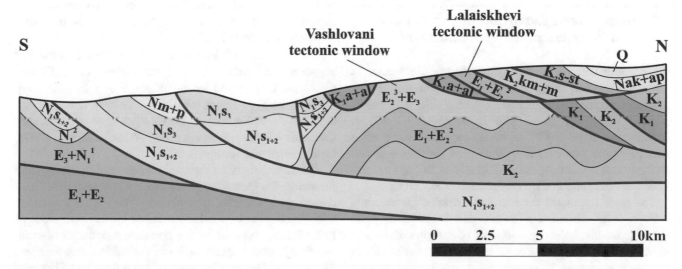

Fig. 2.9 The geological structure of the Alisisgori-Chinchvelta nappe within Gombori ridge (near the cross-section A-A[1] in Fig. 2.2). Stratigraphic signs see below, in Fig. 3.10

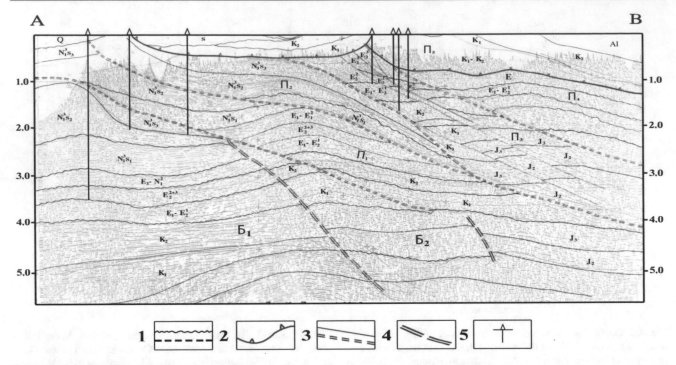

Fig. 2.10 RAMCO seismic reflection profile across the eastern part of Gombori ridg interpreted by Nikoloz Gamkrelidze. 1—boundary surfaces of seismostratigraphic horizons, 2—sole of nappes, 3—thrusts, 4—interbloc faults, 5—boreholes; Б$_{1,2}$—tectonic blocks, П$_1$, П$_2$, П$_3$, П$_4$, П$_5$—tectonic slices. Stratigraphic signs see in Fig. 3.10

of a sub-latitudinal Loki anticlinorium, the pre-Jurassic crystalline basement is exposed. The axis of this structure plunges in both western and eastern directions and causes periclinal closure of the sedimentary cover.

The fold and fault systems of the Adjara-Trialeti, Lok-Karabakh, and the Artvini-Bolnisi zones were formed mainly as a result of the manifestation of Late Alpine (Neogene) tectonic movements with the displacement of masses from south to north (Gamkrelidze 1991, 1997b).

2.5 Geodynamic Evolution of the Caucasus and Paleotectonic Reconstructions

The existing data about geological structure, character of sedimentation and magmatism, geology and the age of ophiolites within the Caucasus, side by side with paleomagnetic data and global plate tectonic reconstructions (Stamfli and Borel 2002) allow us to consider the main features of the geodynamic evolution of the Caucasus and adjacent areas (Gamkrelidze 1986, 1997b, 2016; Gamkrelidze and Shengelia 2005; Gamkrelidze et al. 2015).

The most important for reconstruction of geodynamic settings is to establish a nature and location of paleooceanic basins, its active and passive margins. Existence of oceanic realm in the area of the Mediterranean belt in the Neoproterozoic is shown by a number of various global reconstructions. Birth of Prototethys at that time is

also proved by the existence of ophiolites of Late Precambrian age as in its southern periphery (the Anti-Atlas, the Arabian-Nubian shield, the Loki, Murguz and Tsakhkunyats massifs), so in northern periphery (the Alps, Bohemian and Dzirula massifs) (Gamkrelidze and Shengelia 2005).

Location of the suture line of the Proto-Paleotethys Ocean in the Caucasus is suggested between the Black Sea-Central Transcaucasian and Baiburt-Sevanian terranes, i.e. along the northern periphery of contemporary Somkhit-Karabakh zone (subterrane) (Gamkrelidze 1997a). This is proved by geological and paleomagnetic data, which correspond to criteria of paleomagnetic reliability.

The existence of Paleozoic or older oceanic basins is supposed also in the area of the contemporary Greater Caucasus. It is proved by the existence of Paleozoic ophiolites in the Fore Range zone of the Greater Caucasus and the Klichi ophiolite sheet in the Pass sub-zone of the Greater Caucasus Main Range zone. (Gamkrelidze 1997a; Gamkrelidze and Shengelia 2005).

The main stages of regional metamorphism and granite formation are Grenvilian, Cadomian, Early and Late Caledonian and Variscan orogenesises. They were connected with functioning of subduction zones by the both sides of Proto-Paleotethys and by the northern peripheries of comparatively small oceanic basins of the Arkhyz and Southern Slope of the Greater Caucasus. (Gamkrelidze and Shengelia 2005) (Fig. 2.11).

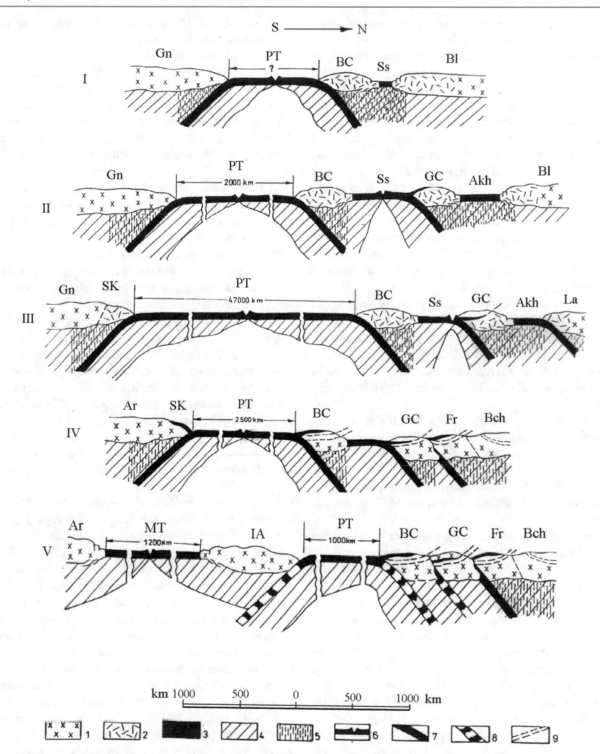

Fig. 2.11 Paleotectonic reconstructions of the Caucasus segment of the Mediterranean orogenic belt along N-S profile for pre-Alpine time (Gamkrelidze and Shengelia 2005). 1. Late Precambrian (Neoproterozoic), II. Late Cambrian, III. Devonian, IV. Early and Middle Carboniferous, V. Late Triassic (the vertical scale is exaggerated by about 10 times). 1—continental crust, 2—subcontinental crust, 3—oceanic crust and obducted ophiolites, 4—upper mantle, 5—stream of heat, fluids and magmatic melts in mantle, 8—Middle oceanic ridge, 7 —subduction zones, 9—surfaces of tectonic layering of Earth crust. Paleooceanic basins: PT-Proto-Paleotethys, Ss—of the Southern slope of the Greater Caucasus, Akh—Arkhys, MT—Mesotetyhis (Neotethis). Continental plates: Gn—Gondvana, Bl—Baltica, La—Laurussia, Ar—Arabia. Terranes: Black Sea—Central Transcaucasian microcontinent GC—Greater Caucasus island arc, IA—Iran Afghan microcontinent. Contemporary tectonic zones: SK—Somkhito-Karabakh, Fr—Forerange, Bch—Bechasin

Considerable tectonic activity in the Caucasus took place during Early Cimmerian (Indosinian) orogeny. In the greater Caucasus with this phase was connected overthrusting of Dizi series to the north, as well as overthrusting of the Elbrus sub-zone infrastructure to the south—into the Pass sub-zone. Early Cimmerian orogeny was manifested in the Dzirula, Loki, Akhum and Asrikchai massifs where it causes a folding of Early-Middle Paleozoic metamorphites and intensive milonitization of the Dzirula massif microcline granites (Gamkrelidze and Shengelia 2005).

Thus Early Cimmerian orogeny, which took place almost in all terranes of the Caucasus, completes formation of the structure of its metamorphic basement.

In the rear of the gradually closing Paleotethys, the joining together of Iran and Arabia and the generation of Neotethys had been taking place already since the Triassic (Gamkrelidze 2016) (Fig. 2.11).

The next extension occurred during the Early Jurassic and beginning of the Middle Jurassic (Fig. 2.12). At the northern active margin of Paleotethys, the Transcaucasus island arc and marginal sea of the Greater Caucasus can be discerned. Within the Lok-Karabach and Transcaucasian island arcs the manifestation of calc-alkaline volcanism of Bajocian age took place.

The next extension occurred during the Early Jurassic and beginning of the Middle Jurassic. At the northern active margin of Paleotethys, the Transcaucasus island arc and marginal sea of the Greater Caucasus can be discerned. Within the Lok-Karabach and Transcaucasian island arcs the manifestation of calc-alkaline volcanism of Bajocian age took place.

One can suppose that the Lesser Caucasus branch or bay of Tethys was formed in the rear of the closure of Paleotethys since the end of the Middle Jurassic. (Gamkrelidze 2016) (Fig. 2.11).

During the Late Cretaceous continues activity of subduction zone along the northern margin of the Lesser Caucasian island arc, which is accompanied by active calc-alkaline volcanism and granitoid magmatism within the Lesser Caucasian island arc replacing to the north by basaltic volcanism of the Adjara-Trialrti intra-arc rift, and then—by alkaline-basaltic volcanism of the Georgian block.

The closure of the Neotethys ocean, as well as of the Paleotethys relic basin, occurred as a result of movements, which spread from north to south. In particular, only the northern part of the Caucasian segment of the Mediterranean belt was affected by the Bathonian (Adygean), Late Cimmerian (pre-Cretaceous), and Austrian (pre-Late Cretaceous) movements. These epochs of tectonic activity are associated with the intense manifestation of andesitic volcanism and granitoid plutonism due to the processes of subduction on the continental margin of the oceanic basins (Gamkrelidze 2016).

The movements of the Austrian phase closed the Lesser Caucasian branch of Mesotethys. At that time, ophiolite nappes were formed in the Lesser Caucasus.

The subsidence that began in the Paleocene reached a maximum in the Eocene, especially in the Middle Eocene, and it was accompanied by calc-alkaline volcanic activity and granitoid magmatism throughout the Lesser Caucasus. Northwards, this was substituted mostly by the accumulation of basaltic sub-alkaline series of the Adjara-Trialeti intraarc rift.

The subsequent phases of Alpine tectogenesis caused the accumulation of molasse deposits, total compression, and final formation of the present-day structure of the Caucasus (Caputo et al. 2000; Gamkrelidze 2016).

The abundance of andesitic and andesite-dacitic volcanism, and granitoid plutonism at the orogenic stage can be related to the continuing activity of subduction zones (intraplate subduction) (Gamkrelidze 2016).

At the same time, maximum compression of the Mediterranean belt caused by an active northward sub-meridional advance of the Arabian Plate and accompanying extensive set of transverse extension fissures was responsible for the penetration of orogenic (collisional) volcanism far into the continent, in a zone of Transcaucasian transversal uplift (Gamkrelidze 2016).

2.6 Summary

The territory of Georgia is a component of the Caucasian segment of the Mediterranean (Alpine-Himalayan) orogenic belt consisting of different rocks whose age varies from Neoproterozoic to Quaternary. The oldest, Precambrian and Lower-Middle Paleozoic mainly metamorphic rocks are exposed in all the tectonic units of Georgia. In particular they are represented within the Main Range zone of the Fold System of the Greater Caucasus, the Transcaucasian Intermountain area (the Dzirula massif), and the Fold System of the Lesser Caucasus (the Khrami and Loki massifs). Mesozoic and Cenozoic formations are developed also in all tectonic units of Georgia and are composed of sedimentary, submarine and continental volcanic and intrusive rocks.

Tectonic structure of pre-Alpine crystalline basement of Georgia is characterized by the development of deep seated tectonic nappes, which are widespread in the Main Range zone of the Greater Caucasus, in the Dzirula and Loki massifs. Alpine structures have a different character in the various tectonic zones. Large southward-directed nappes are developed on the southern slope of the Greater Caucasus.

Geodynamic evolution of the Caucasus are characterized by opening and closing processes of Proto-, Paleo- and Neo-Tethys oceans. The main stages of regional metamorphism and granite formation during Neoproterozoic and

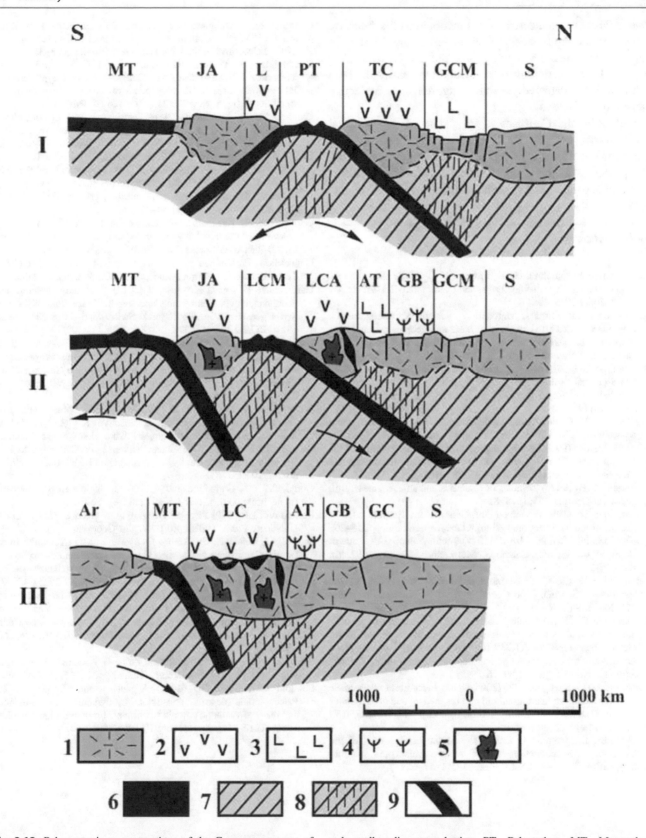

Fig. 2.12 Paleotectonic reconstructions of the Caucasus segment of the Mediterranean orogenic belt along N-S profile for Alpine time (the vertical scale is exaggerated by about 15 times). I—Early Jurassic and beginning of the Middle Jurassic; II—Late Cretaceous; III—Eocene. 1 —consolidated continental crust; 2 to 4—manifestation of volcanism: 2 —calc-alkaline, 3—basaltic, 4—alkaline-basaltic; 5—granitoid magmatism; 6—newly formed oceanic crust and ophiolites; 7—upper mantle; 8—heated upper mantle; 9—subduction zones; Oceanic areas and small sedimentary basins: PT—Paleotethys, MT—Mesotethys (Neotethys), LCM—Lesser Caucasian branch (bay) of Mesotethys, GCM—Greater Caucasus marginal sea. Continental plates and microplates: TC—Transcaucasian island arc, GC—Greater Caucasus island arc, LC—Lesser Caucasus, L—Lok-Karabach Zone, S—Scythian Plate, JA—Iran-Afghanian Plate, LCA—Lesser Caucasian island arc, AT—Adjara-Trialeti intraarc rift, GB—Georgian Block, Ar —Arabian Plate

Paleozoic were connected with functioning of subduction zones by the both sides of Proto-Paleotethys. Later, in Alpine time volcanic activity and granite formation are stipulated by functioning of subduction zones by the both sides of relic Paleotethys and along the northern active margin of Neotethys.

Currently, the Caucasus is in a state of a continental collision between the Arabian plate and Scythian (East European) platform, which leads to current various geological processes in the region: vertical and horizontal movements of separate blocks, earthquakes, volcanic activity etc.

References

Basheleishvili L, Gamkrelidze I, Burtman V (1982) On character of joint of Adjara-Trialeti folded zone and Dzirula massif. Doklady AN SSSR 266(1):196–198

Caputo M, Gamkrelidze I, Malvezzi V, Scrigna V, Shengelaia G, Zilpimiani D (2000) Geostructural basis and geophisical investigations for the seismic hazard assessment and prediction in the Caucasus. Il NuovoCimento 23C(2):191–211

Gamkrelidze IP (1976) Mechanism of tectonic structure formation and some general problems of tectogenesis. Proc Geol Inst Acad Sci Georgia, New Series 52:1–226 (in Russian with extended English summary)

Gamkrelidze IP (1986) Geodynamic evolution of the Caucasus and adjacent areas in Alpine time. Tectonophysics 127(3–4):261–277

Gamkrelidze IP (1991) Tectonic nappes and horizontal layering of the Earth's crust in the Mediterranean belt (Carpathian, Balkanides and Caucasus). Tectonophysics 196(3–4):385–396

Gamkrelidze IP (1997a) Terranes of Caucasus and adjacent areas. Bull Acad Sci Georgia 155(3):75–81

Gamkrelidze IP (1997b) Geologe of Georgia. In: Encyclopedia of European and Asian regional geology. Chapmen and Hal, pp 256–261

Gamkrelidze PD, Gamkrelidze IP (1977) Tectonic nappes of southern slope of the Greater Caucasus. Metsniereba, Tbilisi, 1–87. (in Russian)

Gamkrelidze IP (2000) Once more on the tectonic zoning of the territory of Georgia. Proc Geol Inst Acad Sci Georgia, New Ser 115:204–208 (in Russian)

Gamkrelidze IP (2016) Geological structure of Georgia and Geodynamic evolution of the Caucasus. In: Proceedings of the fourth plenary conference of IGCP 610—from the Caspian to Mediterranean: environmental change and human response during Quaternary, Tbilisi, Georgia, pp 69–76

Gamkrelidze I, Shengelia D (2005) The Precambrian-Palaeozoic regional metamorphism, granitoid and magnetism and geodynamics of the Caucasus. Scientific World, Moscow (in Russian with extended English summary)

Gamkrelidze I, Shengelia D (2007) Pre-Alpine geodynamics of the Caucasus, suprasubduction regional metamorphism and granitoid magmatism. Bull Acad Sci Georgia 175(1):57–65

Gamkrelidze IP, Dumbadze GD, Kekelia MA, Khmaladze II, Khutsishvili OD (1981) Ophiolites of the Dzirula massif and the problem of the Paleotethys in the Caucasus. Geotektonics 5:23–33

Gamkrelidze IP, Dudauri OZ, Nadareishvili GS, Skhirtladze NI, Tutberidze BD, Shengelia DM (2002) Geodynamic typification of Precambrian-phanerozoic magmatism of Georgia. Proc Geol Inst Acad Sci Georgia, New Ser 117:105–126 (In Russian)

Gamkrelidze I, Kakabadze M, Okrostsvaqridze A (2011a) Wide choice of geotraverses and geoparks founding in Georgia. In: Abstract 10th European Geoparks Conference, Langensud, Norway

Gamkrelidze I, Shengelia DM, Chichinadze G, Tsutsunava T, Sun-Lin Chung, Han-Yichiu, Chikhelidze K (2011b) New data on the U-Pb zircon age of the pre-Alpine crystalline basement of the Black Sea—Central Transcaucasian terrane and their geological significance. Bull Acad Sci Georgia 5(1):64–76

Gamkrelidze I, Kakabadze M, Okrostsvaridze A (2012) Kazbegi volcanic region: suitable area for Aspiring the Caucasus, Georgia. In: Abstract of Geoparks 5th UNESCO conference, Japan

Gamkrelidze IP, Pruidze MI, Gamkrelidze MA, Loladze MI (2013) Tectonic map of Georgia (scale 1:500,000). Meridiani, Tbilisi

Gamkrelidze I, Koiava K, Mosar J (2015) Geological structure and geodynamic evolution of the Caucasus. In: Proceeding of the 13th Swiss geoscience meeting, Basel, Switzerland. https://doi.org/10.13140/RG.2.2.19829.68328

Gamkrelidze I, Okrostsvaridze A, Maisadze F, Basheleisvili L, Boichenko G, Skhirtladze I (2019) Main features of geological structure and geotourism potential of Georgia, the Caucasus. Mod Environ Sci Eng 5(5):422–442. https://doi.org/10.15341/mese (2333-2581)/05.05.2019/010

Gamkrelidze I, Shengelia D, Chichinadze G, Yuan His Lee, Okrostavaridze A, Beridze G, Vardanashvili K (2020) U–Pb LA–ICP–MS dating of zoned zircons from the Greater Caucasus pre-Alpine crystalline basement: evidence for Cadomian to Late Variscan evolution. Geologica Carpathica 71(3):240–263. https://doi.org/10.31577/GeolCarp.71.3.4

Gudjabidze GE, Gamkrelidze IP (2003) Geological map of Georgia (scale 1:500,000) (Editor I. Gamkrelidze), Tbilisi

Maruashvili LJ (1970) Physical Geography of Georgia. Part 2, Tbilisi University Press, Tbilisi, pp. 1-347. (in Georgian)

Mauvilly J, Koiava K, Gamkrelidze I, Mosar J (2015) Tectonics in the Greater Caucasus: a N-S section along the Georgian military road—Georgia. In: Proceedings of the 13th Swiss geoscience meeting, Basel, Switzerland. https://doi.org/10.13140/RG.2.2.33251.45607

Mauvilly J, Koiava K, Gamkrelidze I, Mosar J (2016) Tectonics in the Georgian Greater Caucasus: a structural cross-section in an inverted rifted basin setting. In: Proceedings of the 14th Swiss geoscience meeting, Geneva, Switzerland. https://doi.org/10.13140/rg.2.2.26540.56963

Skhirtladze N (1958) Postpaleogene effusive volcanism of Georgia. Publishing House "Mecniereba", Tbilisi, pp. 1–165. (in Russian)

Stampfli GM, Borel GD (2002) A plate tectonic model for the Paleozoic and Mesozoic constrained by dynamic plate boundaries and restored synthetic oceanic isochrones. Earth and Planet Sci Lett 196(1–2):17–33. https://doi.org/10.1016/S0012-821X(01)00588-X

Below three main geotourist routes of Georgia: Tbilisi-Pasanauri-Kazbegi, Tbilisi-Zugdidi-Mestia and Tbilisi-Khashuri-Vardzia are described. Routes are characterized here based on the work of Gamkrelidze et al. 2019 with some changes and additions.

3.1 Tbilisi-Pasanauri-Kazbegi (Along the Georgian Military Road)

The route longwise 155 km crosses eastern termination of the Adjara-Trialeti folded zone, Kura intermountain depression and the Fold system of the Greater Caucasus (Gamkrelidze and Kandelaki 1984; Mauvilly et al. 2015, 2016; Gamkrelidze et al. 2010, 2019) (Fig. 2.2).

3.1.1 Environs of Tbilisi

Acquaintance with the geological section begins in Tbilisi—the capital of Georgia.

Archaeological excavations confirm that the territory of Tbilisi was inhabited as early as BC in the IV millennium. Tbilisi was founded in the fifth century AD by Vakhtang Gorgasali, King of Georgia. It lies on the banks of Mtkvari (Kura) river. Its population is approximately 1.2 million people.

The history of Tbilisi dates back to the sixteenth century. He has been threatened by the enemy many times over the centuries. It has long been conquered by the Persians, Ottomans, Byzantines and Russians. As Tbilisi gathered all languages and religions, it assimilated a different culture, enriched itself and Georgia with diversity. Tbilisi was and is the only city in the Caucasus where you can see the mosque, the synagogue and the church side by side. Therefore, rightly, Tbilisi was the center of the Caucasus and sometimes even the capital.

Old Tbilisi includes the following districts: Abanotubani-Kharpukhi, Kala, Isan-Avlabari, Sololaki, Mtatsminda, Vere, Ortachala, Chugureti, Didube and Nadzaladevi. Old Tbilisi is home to most of the sights of Tbilisi and, consequently, is the main center of tourist attraction.

The ancient Abanotubani District of the old Tbilisi sits below the imposing fortress; the brick, domed rooftops of the baths bubbling up like the water itself. The district is the most historic part of the city, as according to legend, it was the sulphur springs that enticed King Vakhtang Gorgasali to settle the land and declare it the new capital city.

Fast-forward to today: there are five surviving bathhouses in the Abanotubani district where locals and travelers can experience a sulphur bath (Fig. 3.1).

Notable tourist destinations of Tbilisi include cathedrals Sameba and Sioni, Freedom Square, Rustaveli and Agmashenebely avenues, medieval Narikala Fortress, Georgian National Museum and the pseudo-Moorish Opera and Ballet Theater.

Tbilisi is interesting in terms of geological structure as well (Gamkrelidze 1984). The town is surrounded by mountains on three sides. From the West Tbilisi is locked in by the Makhata hills and from the South by the Sololaki ridge and Mount Thabori. The mountain is lowering to the North, from where in sunny weather the Caucasus Range with its majestic Kazbegi is distinctly seen. In the North-West end small lakes are situated in the lows of the mountain-relief-Kustba (Turtle Lake) and Lisi and behind the Makhata ridge the Tbilisi Sea—an artificial reservoir is located In the vicinity of Tbilisi, on the right bank of the Kura the following rather gentle folds of latitudinal strike are distinguished (from south to north): Teleti anticline, Krtsanisi syncline, Mamadaviti anticline, Saburtalo syncline, Lisi anticline, Digomi syncline and Mtskheta anticline belonging to the eastern termination of the Adjara-Trialeti zone (Gamkrelidze et al. 2010) (Fig. 3.2).

I. Gamkrelidze et al., *Geotourism Potential of Georgia, the Caucasus*, Geoheritage, Geoparks and Geotourism, https://doi.org/10.1007/978-3-030-62966-3_3

Fig. 3.1 The view of the old Tbilisi (photography by Ilan Molcho from Israel)

The neighbourhood of Tbilisi is mainly built up outcrop the oldest Middle Eocene rocks. To the East of the Kura these deposits subside beneath younger formations (Fig. 3.2).

Stop 1. The participants will begin the acquaintance with Tbilisi and the geological structure of its neighbourhood from the top of the Mtatsminda (Saint Mountain, or as it is called "Mamadaviti (Father David)"). On the upper plateau of Mtatsminda (727 above sea level) a park is situated. The traditional way to get on the plateau is the Funicular, built in 1905. From the terrace and special places made in different parts of the park a magnificent view of Tbilisi opens up.

Stop 2. The exposure of the Metekhi cliff on the left bank of the Kura is built up of the olistostrome strata of the Middle Eocene confined to the perikline of the Mamadaviti anticline (Gamkrelidze et al. 2010).

In the Metekhi cliff crop out above mentioned Tbilisi olistostromes of Middle Eocene age, which first was described by H. Abich as "conglomerates of entangled bedding" (Gamkrelidze et al. 2010).

Above of the Metekhi cliff in the thirteenth century was built Metekhi church. There stands the monument of Vakhtang Gorgasali—founder of Tbilisi (Fig. 3.3).

South of the Metekhi cliff on the sky-line is seen the Teleti range lowering to the East. It coincides with the southernmost Teleti anticline of the Adjara-Trialeti zone. Relatively narrow Krtsanisi syncline is situated between the Teleti and Mamadaviti anticline, formed of the Middle and Upper Eocene deposits (Fig. 3.2).

The olistostrome rocks are underlain by thick-bedded tuff sandstones, argillites and marls of the Middle Eocene (Dabakhana beds).

Mainly in the lower part of the Tbilisi olistostromes typical submarine landslide folding is widely developed.

Stop 3. Along the Tbilisi-Rustavi road, submarine land-slide isoclinal folds of Middle Eocene tuff sandstones, which are located between undislocated layers of the same rocks, are clearly visible (Fig. 3.4).

In the northern limb of Mamadaviti anticline the Middle Eocene is superimposed by the Upper Eocene terrigenous gypsiferous series with thick coarse-grained sandstones containing pebbles of the Middle Eocene rocks. The lower part of these series is represented by thin-bedded sandstones, shales and marls with fish scales, about 100 m thick. Higher this suite is replaced by the Tbilisi suite (Nummulite suite)—alteration of shales, aleuorolites, marls and sandstones with redeposited tuffaceous material.

Tbilisi olistostromes are built up of different size mostly angular and, in places, more or less rounded inclusions of Cretaceous limestones and effusive rocks (andesites, basalts, rhyolites dacites) typical for Adjara-Trialeti zone, as well as terrigenous rocks (sandstone-siltstone and clastic limestone turbidites, tephroturbidites, pelagic argillites, motley clays, marls and tuffs) representing, on the basis of all signs, fragments of Paleocene-Lower Eocene flysch and of the Lower part of the Middle Eocene of Adjara-Trialeti zone. Consequently, all these fragments represent so-called

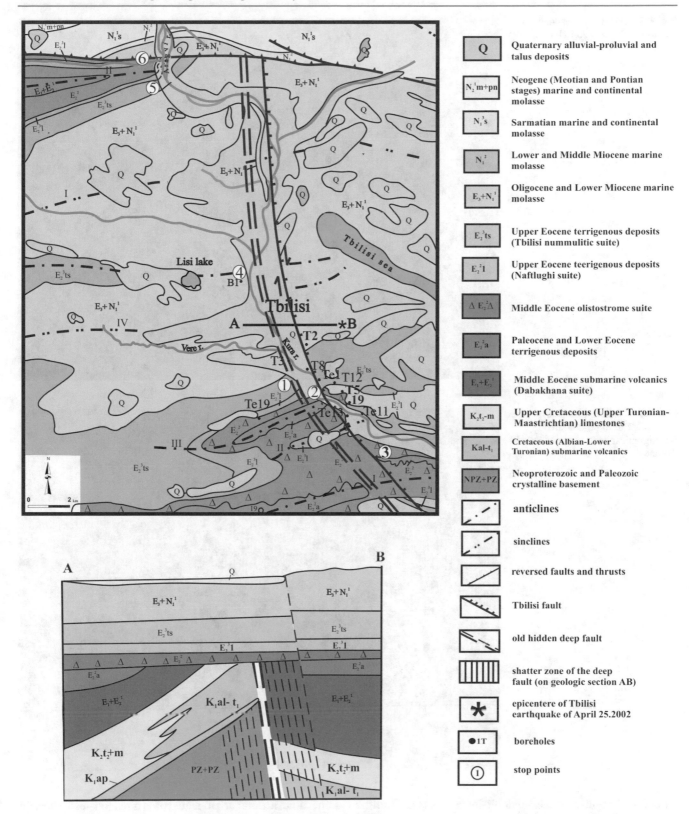

Fig. 3.2 Geological map of the eastern termination of the Adjara-Trialeti zone in vicinity of Tbilisi. Clipping from the Geological map (1:500,000) of Georgia (Gudjabidze and Gamkrelidze 2003) modified; Folds (in roman numerals): I—Teleti anticline, II—Krtsanisi syncline, III—Mamadaviti anticline, IV—Saburtalo syncline, V—Lisi anticline, VI—Digomi syncline, VI—Mtskheta anticline

Fig. 3.3 Tbilisi olistostromes outcrop under Metekhi church

Fig. 3.4 Submarine landslide lying folds in multi-coloured inequigranular tuffs of the Middle Eocene

intraclasts but olisostromes-endoolistostromes (Gamkrelidze 1984; Gamkrelidze et al. 2008, 2010).

These data undoubtedly indicate that source of removal of clastic material was within the Adjara-Trialeti zone. At the same time a number of signs of Tbilisi olistostromes, in particular, the presence of erratic masses from underlying deposits of the same basin, relatively small thickness, the absence in them of tectonically reworked particles, availability of clearly expressed slump-downfall structures and textures, primary normal interrelations of olistostromes with enclosing deposits, the presence of normal-sedimentary rocks interbedding with olistostromes, undoubtedly concern these formations to typical gravitational olistostromes (Gamkrelidze 1984; Gamkrelidze et al. 2008, 2010).

Analysis of boring data and field observations allow approximately outline Tbilisi olistostromes spreading area, occupying about 500 km^2. Besides, these data indicate that the presence of washing out land and considerable scarp in relief, which was necessary for the formation of Middle Eocene olistostromes, and corresponding submeridional deep fault, one can assume only between the Tbilisi boreholes № № 1, 2, 3 in the east and the borehole Tel-19 in the west (Fig. 3.2) (Gamkrelidze et al. 2008, 2010).

Taking into account peculiarities of geological structure of the eastern part of the Adjara-Trialeti zone and borehole data, one can assume that generation of this old deep normal fault took place seemingly already at the close of the Early Cretaceous and it continued to exist during the accumulation of thick Albian-Lower Turonian volcanogenic suite, which developed mainly in eastern lying side of this fault (normal fault), but in hanging wall it replaced by carbonaceous deposits. The fault developed or underwent rejuvenation in the second half of the Middle Eocene (Gamkrelidze et al. 2008, 2010).

Formation of scarp in relief and coastal land was connected apparently with manifestation of Trialetian orogeny in the second half of the Middle Eocene when all this highly elevated land has been completely disintegrated, washed out and overlain by olistostrome stratum, which is built up of clastic material of this land constituent rocks, and then—by the transgressive Upper Eocene (Gamkrelidze et al. 2008).

It should also be noted that on the bases of geomorphic, hydrogeologic and seismic data, as well as distribution of rocks' temperature field very close to this fault, along the Kura River very young seismogenerating fault is established, which arose close to old fault, at the boundary between the thick shatter zone, that is typical for such deep faults, and the monolithic rocks (Gamkrelidze et al. 2008, 2010).

For the purpose of revealing of expressed in topography active fault within the Tbilisi city satellite image (Shuttle Radar Topography Mission) of this region have been studied. Study of satellite image (Fig. 3.5) reveals the presence of clearly rectilinear scarp in relief with the height, directly in the scarp, 15–20 m. The scarp well fixed between the environs of Grmagele in the north up to the Metekhi church in the south and perpendicularly cuts strike of folds of eastern termination of Adjara-Trialeti zone (Digomi syncline, Lisi anticline, Saburtalo syncline and Mamadaviti anticline) and lithologically different Oligocene-Lower Miocene, Upper Eocene and Middle Eocene deposits (Gamkrelidze et al. 2008, 2010).

Satellite ASTER image data (Thermal Infrared Range) gave us information about distribution of temperature field in the rocks constituting of Tbilisi vicinity and adjacent areas (Fig. 3.6). Heterogeneity of distribution of temperature field is due to different features and structures of objects. On the satellite image is distinctly observed very sharp border between the Lower and the relatively higher temperature area completely coinciding with above mentioned rectilinear scarp in relief (Gamkrelidze et al. 2008, 2010).

Aforesaid, in our opinion, indicates the existence in this part of the city of steep fault with average strike azimuth 3350, north-eastern wall of which is uplifted. To the north-west the fault continues along the Kura valley apparently up to Northern marginal deep fault of Adjara-Trialeti zone, but to the south-east, probably along the right bank of the Kura River, in thick alluvial-proluvial deposits towards Rustavi town (Fig. 3.2). Taking into account the presence of thick (35–40 m) alluvial deposits of the Kura River in lying wall of the fault (borehole Tbilisi-4), one can suppose that its summary vertical separation in its central part reaches 80–100 m. Another sign of this young fault existence is sharp change of the Kura stream direction from sublatitudinal to submeridional in this segment of the river and its antecedent character (Gamkrelidze et al. 2008, 2010).

Young Tbilisi fault apparently arose close to above mentioned old fault before the Pliocene (Rodanian orogeny) or mainly before the Pleistocene (Wallachian orogeny) at the boundary between the thick shatter zone, which is typical for such deep faults, and the monolithic rocks and continues to develop at present (Fig. 3.2). Consequently in fact here took place a rejuvenation of old deep normal fault at neotectonic stage but with reversed sign of displacement that is typical for many faults of the territory of Georgia.

Thus, there need be no doubt that Tbilisi earthquake of April 2002 was concerned exactly with this—Tbilisi fault as well as historical earthquakes in 1682, 1803, 1804, 1819. These were moderate events with intensity 5–6.

Judging by position of Tbilisi 2002 earthquake epicenter (see Fig. 3.6), as well as by depth of its hypocenter (7.3 km), fault plane must dip to the north-east approximately at an angle of 70°.

After getting acquainted with the geology of the outskirts of Tbilisi, the excursion goes to the north along the Georgian military road.

Stop. 4. Theroute of the excursion passes through the Saburtalo syncline consisting of clayey-sandy deposits of the Maikop series of the Oligocene. After leaving the city and covering a distance of 1.5 km the rout crosses the Lisi anticline, formed by the lower part of the Oligocene, and then the wide syncline Digomi Valley. The syncline consists of the upper part of the Upper Eocene-Akhalsopeli suite

Fig. 3.5 SRTM—Shuttle Radar
Topography Mission. Spatial
Resolution = 90 v.
Wavelength = 5.6 and 3.1 sm.
Data: 2000, 17.02. (Gamkrelidze
et al. 2008, 2010)

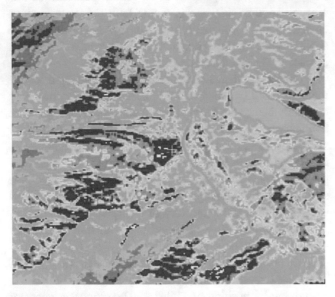

Fig. 3.6 Distribution of Temperature field (Celsius) in the rocks from
ASTER image 13 TIR band. Wavelength = 10.6 m. Special Resolution
TIR = 90 m. Data: 2002, 31. 01. (yellow—reddish colours represent
warm areas, while green to blue colours correspond to cold areas).
(Gamkrelidze et al. 2008, 2010)

(alternation of green sandstones and clays) and of the thick
(up to 300 m) clay-sandy Maikop series of the
Oligocene-Lower Miocene (Gamkrelidze et al. 2010).

Stop 5. The Maikop series is usually represented by thin
bedded gypsiferous clays with fish scales, large septarian
concretions and jarosite coatings. However, in East Georgia
a facies of coarse-grained quartz sandstones (Lower Miocene) is often related to the tops of the Maikop series.
A thick (up to several hundred meters) suite of such

sandstone outcrops in the northern limb of the Digomi
syncline forming continuous scarp along the right bank of
the Kura. The appurtenance of these sandstones to the
Sakaraulian regiostage (Burdigalian) has been proved by
means of microfauna. These sandstones are overlain by
upper clayey part of this series, occurring in the core of the
syncline and belongs to the Kotsakhurian regiostage (Hel-
vetian), while the clays, underlying the sandstones, are the
Oligocene part of the Maikop series (Gamkrelidze and
Kandelaki 1984; Gamkrelidze et al. 2010, 2019).

The lower part of the Upper Eocene is represented by
foraminiferal marls of the Lirolepis horizon with sandstone
interlayers dipping south. The Upper Eocene sandstones are
grouped into separate packets and form small ridges. On one
of these packets the power station dam is built and on the left
bank of the Kura, opposite the town of Mtskheta, the mar-
velous monument of the ancient Georgian architecture rises
—the Jvari convent (convent of the cross) (Fig. 3.7). It was
founded in 586–600 years BC on the place where, as legend
has it, a large wooden cross has stood since the fourth
century, when Christianity was adopted by Georgia as an
official state religion. Mtskheta is one of the oldest cities in
Georgia. Up to sixth century AD it was the capital of the
Georgian kingdom. The city is extremely rich in ancient
monuments. In the center of the city stands a magnificent
building of the first quarter of the eleventh century—
Svetitskhoveli Cathedral (Fig. 3.8).

Stop 6. The outskirts of Mtskheta are interesting from the
geological viewpoint as well. Here at the confluence of the
Kura and Aragvi rivers two geotectonic units collide: the
Adjara-Trialeti folded zone and the Georgian block (Kartli

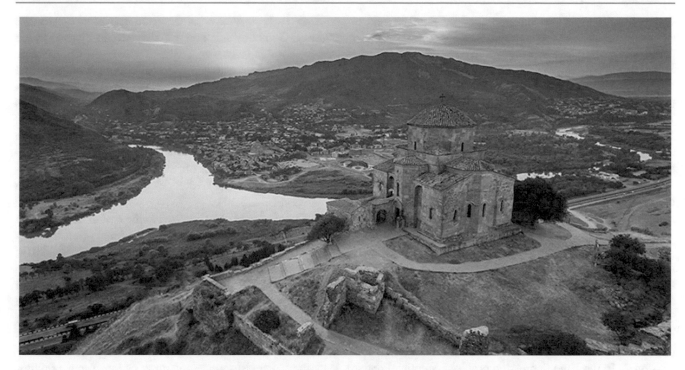

Fig. 3.7 Jvari convent (photography by Giorgi Shermazanashvili)

Fig. 3.8 The Svetitskhoveli
Cathedral (photography by Levan
Gakadze)

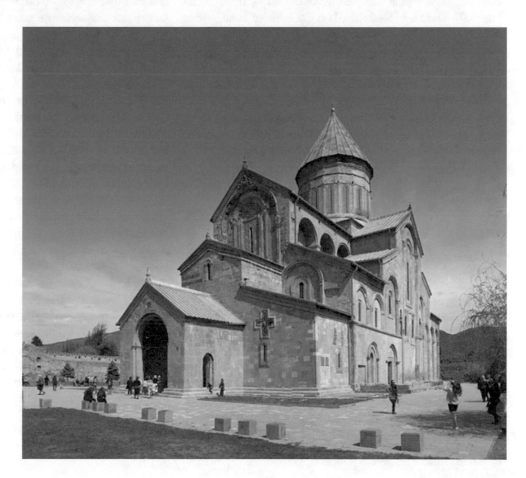

depression). The former is represented by the Mtskheta or Armazi anticline, whose northern limb is complicated by additional folding. In the core of the fold the Paleocene-Lower Eocene deposits outcrop (1000 m), while the limbs consist of the Middle Eocene volcanic formation (500 m) and the Upper Eocene clayey-sandy rocks (800 m). The fold axis, traced in the latitudinal direction along the Kura, gradually rises westward, and at the village of Dzegvi, in its core, carbonaceous deposits of the Upper Cretaceous appear (500 m). In this part the Mtskheta anticline is thrust over the Georgian block. In their turn Miocene deposits of the Georgian block are overthrust upon the Adjara-Trialeti zone (Gamkrelidze and Kandelaki 1984; Gamkrelidze et al. 2010, 2019).

Beyond Mtskheta overthrust on the left bank of the Kura the monoclinal ridge Skhaltba is seen, which consists of the following deposits: Maikop clays and sandstones of the Oligocene-Middle Miocene (1400 m), Lower and Middle Sarmatian (50 m), continental sandstones and clays of the Upper Sarmatian (1500 m), continental Lower Pliocene (Meotian-Pontian) conglomerates (2000 m).

Stop 7. In the Kura depression crop out Meotian-Pontian conglomerates, which completely composed by flysch rocks (Fig. 3.9). Their age established on the basis of stratigraphic position and vertebrates of Meotian age (Meladze 1967; Gamkrelidze and Kandelaki 1984; Gamkrelidze et al. 2010, 2019).

In the Kura depression according to seismic reflection profiling and seismic tomography many thrusts are developed, which at the depth become subhorizontal. They arose under the influence of nappes of the southern slope of the Greater Caucasus (Gamkrelidze and Gamkrelidze 1977) (Fig. 3.10).

Fig. 3.9 Meotian-Pontian conglomerates (Dusheti suite) near v. Bodorna

Stop 8. Near the v. of Zhinvali the route enters the Zhinvali-Gombori subzone of the Mestia-Tianeti flysch zone, within which the Alisisgori-Chinchvelta, Sadzeguri-Shakhvetila and Zhinvali-Pkhoveli nappes and the Ksani-Arkala parautochthone are distinguished, representing in geological past independent structural-facies zones.

Stop 9. To the north from the Kura depression the route crosses the Ksani-Arkala parautochthon, which is a comparatively thin plate torn off the autochthone (eastern continuation of the Gagra-Java zone) and thrust upon the Kura depression (Fig. 3.10).

Different nappes crop out well to the east of the Georgian Military Road, along the river Pshavis-Aragvi (left tributary of the river Aragvi) (Gamkrelidze et al. 2010, 2019; Gamkrelidze and Maisadze 2016).

Stop 10. Along the river Pshavis-Aragvi, in the frontal part of the Sadzeguri-Shakhvetila nappe, one can observe a thin slice consisting of intensively dislocated Paleogene deposits, which form a heavily compressed syncline overturned to the West-South-West. Along the highway dark-grey, greenish-grey, sometimes carbonaceous argillites and schistose clays crop out with interlayers of carbonaceous sandstones and schistose marls of the Paleocene Shakhvetila suite, as well as greenish-grey and green schistose marls, carbonaceous argillites, and sandstones of the Kvakevris-khevi suite of the Lower-Middle Eocene (Figs. 3.10 and 3.11). Along the r. Pshavis Aragvi in Upper Cretaceous and Paleogene rocks intensive minor folding is observed (Fig. 3.12). To the east, on the left slope of the river Pshavis Aragvi, along the Zhinvali-Tianeti highway, a synclinal composed of Paleogene carbonate mudstones and sandstones is well exposed, in which also intensive minor folding is developed (Fig. 3.13).

To the North-East the Alisisgori-Chinchvelta nappe is observed, whose lower part consists of the terrigenous Navtiskhevi (Albian), Ukughmarti (Lower Cenomanian), Ananuri (Upper Cenomanian-Lower Turonian) and carbonaceous Margalitisklde (Upper Turonian) and Eshmakiskhevi (Coniacian-Santonian) suites. Along the highway is observed an overthrusting of flat-lying dark-grey, greenish, thin-bedded argillites and medium and thin bedded grey-rusty grey sandstones of the Lower Cenomanian (Ukughmarti suite) and sometimes sillicites of the Upper Cenomanian-Lower Turonian (Ananuri suite) of the Alisisgori-Chinchvelta nappe upon the heterochronous rocks of the Sadzeguri-Shakhvetila nappe dipping at high-angles. Both nappes are ruptured by young sublatitudinal reverse faults seen along the route of the excursion (Fig. 3.11) (Gamkrelidze and Kandelaki 1984; Gamkrelidze et al. 2010, 2019).

Stop 11. Returning to the gorge of the river Aragvi, along the road one can see Upper Eocene olistostromes of the southern slop of the Greater Caucasus.

The olistostromes consist of olistoliths of the Upper Jurassic reef limestones, volcanic rocks of the Bajocian, shales and sandstones of the Lias and various crystalline rocks of the Paleozoic, cemented by sandy-clayey deposits (Figs. 3.14 and 3.15). That is source of these rocks was the Gagra-Java zone of the Greater Caucasus (so-called Racha-Vandam Cordillera), which at present is entirely overlapped by nappes of the Flysch zone in the eastern segment of the southern slope (east of the Rioni River). This cordillera was a major source of clastic material for Upper Cretaceous and especially abundant for Upper Eocene sediments (Maisadze 2008, 1994).

From the South-West the strip of the Upper Eocene olistostromes outcrops is bounded by the large Zhinvali-Pkhoveli overthrust overlapping the Ksani-Arkala paraautochthone (Figs. 3.10 and 3.11).

By the end of the Late Eocene, at the maximum of the Pyrenean tectonic phase, landslide events were abundant due to the beginning of horizontal displacements and catastrophic earthquakes (Maisadze 2008).

The Pyrenean movements in that period were related to displacement of the Transcaucasian block to the north and its underthrusting beneath the Fold system of the Greater Caucasus, which was most evidently expressed in the major (pre—Late Pliocene—Rodanian) phase of the nappe formation (Gamkrelidze and Gamkrelidze 1977).

Stop 12. Further, near the Ananuri fortress one can see that the Zhinvali Pkhoveli nappe is represented here by one large Ananuri anticline. (Figs. 3.10 and 3.11). The core of the fold is built up with Aptian shales and sandstones (Tetrakhevi suite) and variegated marls and sandstones of Albian age (Navtiskhevi suite). On vertical southern limb of Ananuri anticline Ananuri fortress, one of the best monuments of architecture of late feudalism is situated. Limbs of the fold are composed of sandstones and thin-laminated marls of Lower Cenomanian (Ukughmarti suite), Upper Cenomanian-Lower Turonian silicified thin laminated marls and brown silicates (Ananuri suite), red marls, lithographic limestones of the Upper Turonian-Santonian (Margalitisklde and Eshmakiskhevi suites) Maastrichtian deposits are represented by orbitoid thin-laminated marls and limestones with interlayers of microconglomerates, gradually continuing by marls of Danian age. Rather to the east of this locality in the vicinity of the village Zhinvali Maastrichtian deposits (Orbitoide suite) are almost completely composed of olistostromes (Gamkrelidze and Kandelaki 1984; Gamkrelidze et al. 2010, 2019).

Not far from here, on the left slope of the Arkala river (left tributary of the Aragvi river), along the Transcaucasian gas pipeline one can observe outcrops of root zone of the retro-overthrust, which is characterized by evident squeezing of olistostromes, their thrust to the north and displacement within the flysch sequence (Fig. 3.11). This root zone in the paper by Gamkrelidze and Maisadze (2016) is described in detail.

Stop 13. From here begins the Sadzeguri-Shakhvetila nappe where Paleogene deposits in sandstone-siltstone flysch facies and often interruptions in sedimentation (before the Albian, Cenomanian, Maastrichtian and Paleocene) are wide developed (Gamkrelidze and Gamkrelidze 1977).

Stop 14. From the v. of Pavleuri begins Utsera-Pavleuri nappe, in which from North to south smaller imbricated slices are distinguished, built up of clays and sandstones of the Lower Cretaceous (Aptian-Albian), siliceous-sandy deposits (Cenomanian) and limestones of the Turonian-Lower Santonian (Fig. 3.16).

Stop 15. Clay shales of the Aptian (Dgnali suite) near the v. of Bibliaani overthrust various horizons of the Upper Cretaceous clastic-limestone flysch deposits on the so-called Tladon thrust. This fault bounds the Glola-Mleta slice from the South (Fig. 3.16).

The area of Pasanauri is characterized by scenery, climatic and balneological conditions of mountain health-resort. The mineral spring of the Essentuki type has been known here since ancient times. Low flow water (700 lt/day). It is associated with the core of the Pasanauri anticline line, consisting of the Hauterivian-Barremian sand-marl-sand deposits (Bakhani and Pasanauri suites).

Stop 16. Small town of Pasanauri is located at the confluence of the rivers White and Black Aragvi (Fig. 3.17). Between Pasanauri and Gudauti thrusts exists a large Glola-Mleta tectonic slice. It is mainly formed of Lower Cretaceous deposits with minor folding (Fig. 3.18).

Stop 17. Beyond the Gudauri tunnel-passageway the road descends along the andesite-basalte Gudauri lava flow of the Sakokhe-Sadzele volcano to the v. of Kvesheti (Fig. 3.19). This lava fills the paleoriver channels of the White Aragvi and its left tributaries.

Gudauri is young and rapidly developing winter sports resort located at Kazbegi region of Georgia, 120 km from Tbilisi, at the height of 2100 m (about 7200 ft) near the Cross Pass. Skiing season lasts from December to April, comfort skiing on all routes. In May skiing is possible on the five (highest) lifts and even on the second lift in snowy years.

Stop 18. The Jvari (Cross) Pass affords a panorama of the mountain-and-valley topography of the Aragvi river headwaters. To the west of the Pass the Keli volcanic upland can be seen built up of a group of extinct Pleistocene-Holocene volcanoes termed Seven Brothers. Fragments of lava flows from these volcanoes are preserved in the area of the Cross Pass. At present they are separated from their sources by a deep canyon (400–500) of the White Aragvi river. Small

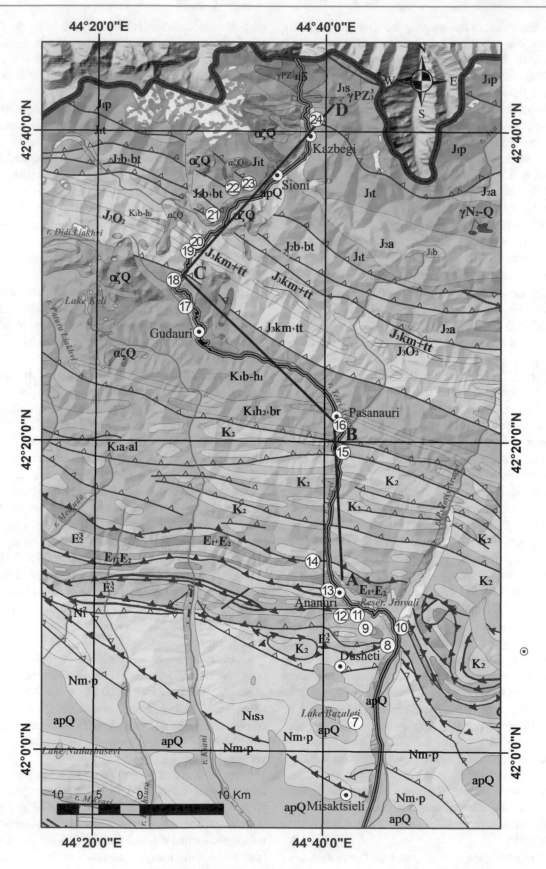

◀ **Fig. 3.10** Geotourist route Tbilisi-Pasanauri-Kazbegi. (along the Georgian Military Road). Clipping from the Geological map (1:500,000) of Georgia (Gudjabidze and Gamkrelidze 2003) showing the geological structure along the route and stop points (numerals in circles). A-B-C-D-segmented line of geological profile. The meaning of stratigraphic signs along the route: Q—Quaternary, $\upsilon\zeta Q$—Upper Pliocene-Lower Quaternary continental lavas, $N_2ak + ap$—Akchagilian and Apsheronian, $Nm + p$—Meotian and Pontian, N_1s_{1+2}—Lower and Middle Sarmatian, N_1s—Sarmatian, N_1^2—Middle Miocene, $E_2^3 + E_3$—Upper Eocene-Oligocene, E_2^3—Upper Eocene, E_2—Middle Eocene, $E_1 + E_2$—Paleocene and Eocene, $E_1 + E_2^1$—Paleocene and Lower Eocene, K_2—Upper Cretaceous (undismembered), $K_2km + m$—Campanian-Maastrichtian, K_2t_2-m—Upper Turonian-Maastrichtian, K_2s-st—Cenomanian-Santonian, K_1al-K_2t_1—Albian and Lower Turonian, K_1a-al—Aptian and Albian, K_1h_2-br—Upper Hauterivian-Barremian, K_1b-br—Beriasian-Barremian, J_3—Upper Jurassic (undismembered), J_2bt—Batonian, J_2b—Bajocian, J_1-J_2a—Lower Jurassic-Aalenian, J_2a—Aalenian, J_1t—Toarcian, J_1P—Plinsbachian, J_1s—Sinemurian, gPZ_3^2—Late Variscan granitoid intrusives.

remnants of flows occur in the vicinity of the v. Khatissopeli, on right bank of the river Aragvi (Gamkrelidze and Kandelaki 1984; Gamkrelidze et al. 2010, 2019).

The band of Upper Jurassic clastic-limestone flysch deposits, between the Gudauri and Truso thrusts, represents a large Kolosani-Pakhviji tectonic slice, which overlaps a series of folds of the Lower Cretaceous carbonaceous rocks (Fig. 3.20).

Stop 19. In the middle course of the Bidara gorge (right tributary of the Terek), at the slope of the gorge some fragments of Quaternary andesitic agglomerate tuffs and lava breccias of the Sakokhe-Sadzele volcano are presented. Numerous ferrous-carbonate mineral springs of the "Narzan" type flow out from fissures of carbonaceous rocks. Sometimes large accumulation of red and yellow-brown carbonaceous tuffs (travertines) is observed around the springs.

Stop 20. South of the v. Kobi—along the Bidara gorge one can see the Upper Jurassic rocks consisting of clastic-limestone flysch deposits. They crop out without interruption up to the Cross Pass and form numerous small narrow folds overturned southward.

Stop 21. Near the v. Kobi on the left bank of the Terek under the Mnadoni andesite lava clay-shales, quartz and polimictic sandstones and rarely sandy–limastone Shevardeni suite of the Bathonian age (250–300 m) are exposed. The suite is thrust southwards onto various horizons of the Upper Jurassic clastic-limestone flysch. From this thrust begins the Kazbegi-Lagodekhi zone of the Greater Caucasus (Fig. 3.20).

Stop 22. Between villages Kobi and Sioni a large volcanic edifice of the mount Kabarjina (3121 m) occupies the left bank of the Terek (Fig. 3.21). This is a polygenetic stratovolcano, built up of heterogeneous volcanic facies produced by several eruptive phases. At the base of the volcano a thick series of andesite-dacitic agglomerated tuffs and lahar breccias occur; they are overlain by lava of similar composition cut through by dacitic extrusions. The volcanic edifice is dated Middle-Late Pleistocene.

Further at a distance of 6 km along the left slope of the Terek the outcrop is strongly dislocated, broken by imbricated overthrust faults, Upper Aalenian clay shales, aleurolites and sandy limestones of the Upper Ukanapshavi suite (300 m) and Bajocian clay slate of the Mnadoni suite.

Stop 23. In the vicinity of Sioni the Toarcian shaly-sandy deposits are thrust upon the Bajocian clay slates and argillites of the Mnadoni suite along the large Amel-Chaukhi thrust (Fig. 3.20). This zone comprises several eruptive centers of the Quaternary andesite-dacite lava flows filling the beds of the left tributaries of the Terek.

Further to Kazbegi the route crosses the band of complexly dislocated Toarcian rocks. They are complicated by three parallel thrusts, which comprise numerous carbonaceous springs of the Kazbegi type (Gamkrelidze and Kandelaki 1984; Gamkrelidze et al. 2010, 2019; Mauvilly et al. 2015, 2016).

Stop 24. To the north from Kazbegi along the road between the Plinsbachian and Toarcian shaly-sandy deposits one can see the steeply dipping northward Main thrust of the Greater Caucasus limited from south the Main Range zone of the Greater Caucasus (Fig. 3.20) (Gamkrelidze and Kandelaki 1984; Gamkrelidze et al. 2010, 2019).

3.2 Tbilisi-Zugdidi-Mestia-Ushguli

At the beginning participants of the geological excursion get acquainted with the sights and geological structure of environs of the city of Tbilisi, and then to the town of Mtskheta with the geology of the eastern termination of the Adjara-Trialeti folded zone. First 6 stop points are described above (see chapter Tbilisi-Pasanauri-Kazbegi (along the Georgian Military Road) ("Environs of Tbilisi").

The rout longwise 450 km at the beginning crosses the eastern termination of the Adjara-Trialeti zone and further, along the Mtkvari (Kura) river—Neogene terrigenous deposits.

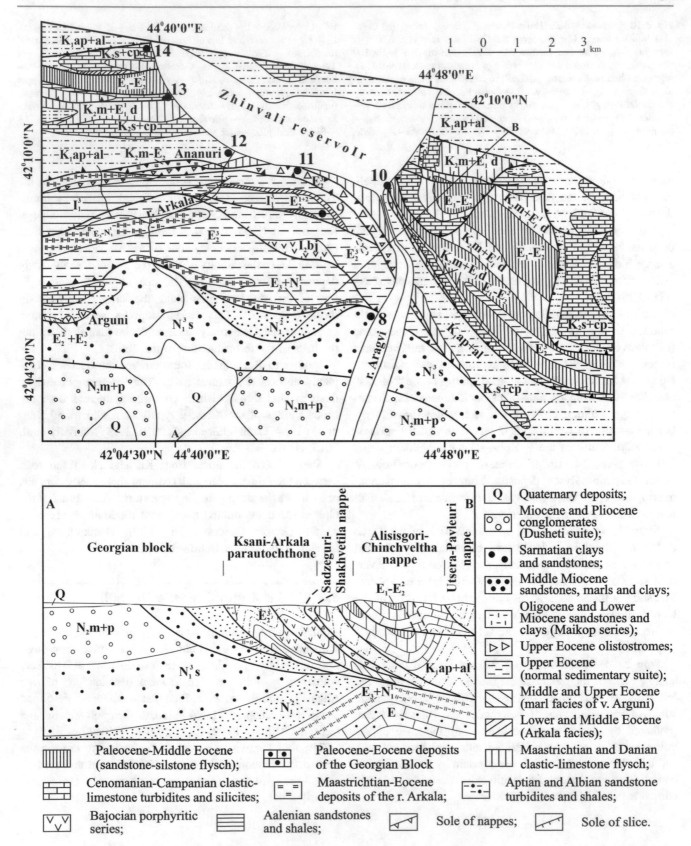

Fig. 3.11 Geological map of the Aragvi river basin (in environs of Zhinvali and Ananuri villages) (according to Gamkrelidze and Maisadze 2016)

Fig. 3.12 Minor folding in Campanian limestones and marls on the left bank of the Pshavis Aragvi river

Fig. 3.13 Minor folding in Paleogene carbonate mudstones and sandstones along the Zhinvali-Tianeti highway

Fig. 3.14 Olistoliths of Bajocian volcanic rock and Upper Jurassic reef limestone in Upper Eocene olistostromes (between Zhinvali and Ananuri) (Gamkrelidze et al. 2010)

Fig. 3.15 Olistolith of Upper Jurassic reef limestone in Upper Eocene olistostromes (between Zhinvali and Ananuri) (Gamkrelidze et al. 2010)

In south-eastern part of the pre-Alpine Dzirula crystalline masiff, in so-called Chorchana-Utslevi zone are preserved the Lower and Middle Paleozoic and Precambrian-Paleozoic metaophiolites (serpentinites and amphibolites) dated by fauna (Figs. 3.22, 3.23 and 3.24).

Earlier these formations, traced at a distance of about 14 km, to the north from villages Chorchana and Utslevi, in the Cheratkhevi and Lopanistskali river basins, were known as "suite of metamorphic schists", also referred to as "phyllite" or "diabase-phyllite stratum."

Fig. 3.16 Cross-section between Pavleuri and Pasanauri (A–B segment of profile) (Gamkrelidze et al. 2019)

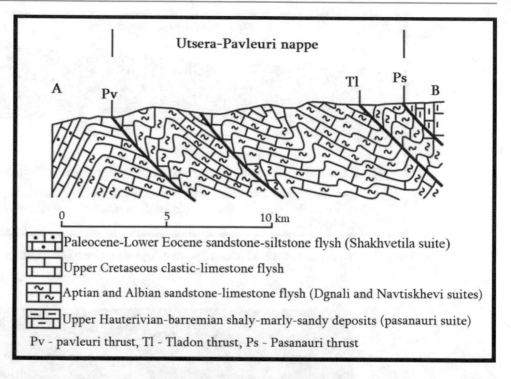

Fig. 3.17 The confluence of the rivers White and Black Aragvi (photography by Giorgi Shermazanashvili)

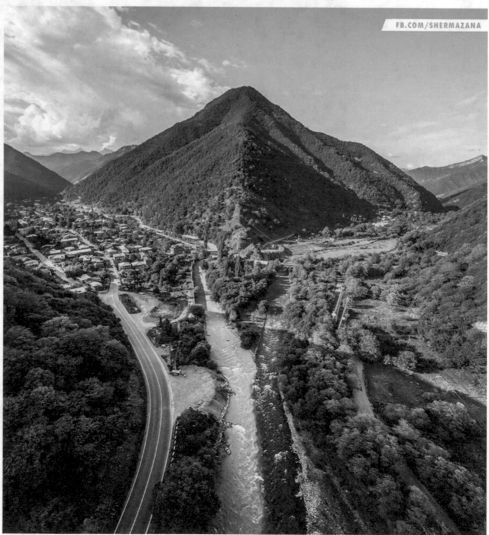

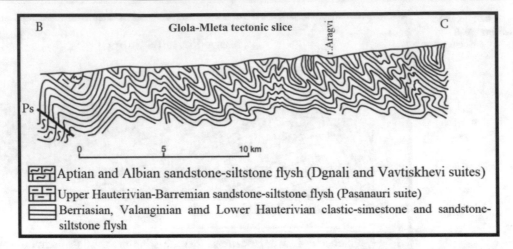

Fig. 3.18 Cross-section between Pasanauri and Jvari (Cross) Pass (B–C segment of profile) (Gamkrelidze et al. 2019)

Fig. 3.19 Columnar jointing in the Gudauri lava flow

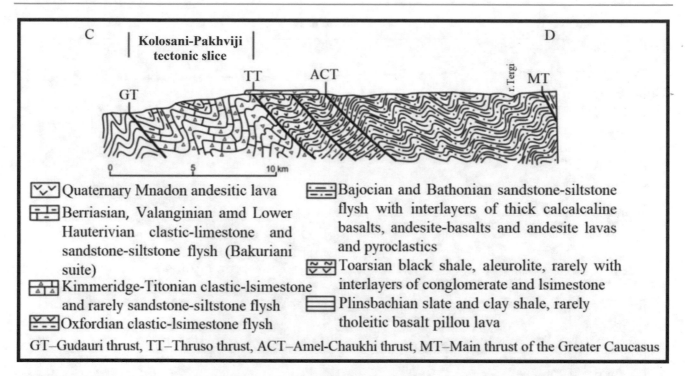

Fig. 3.20 Cross-section between Jvari (Cross) Pass and Kazbegi (C–D segment of profile) (Gamkrelidze et al. 2019)

Fig. 3.21 Kabarjina Quaternary volcano

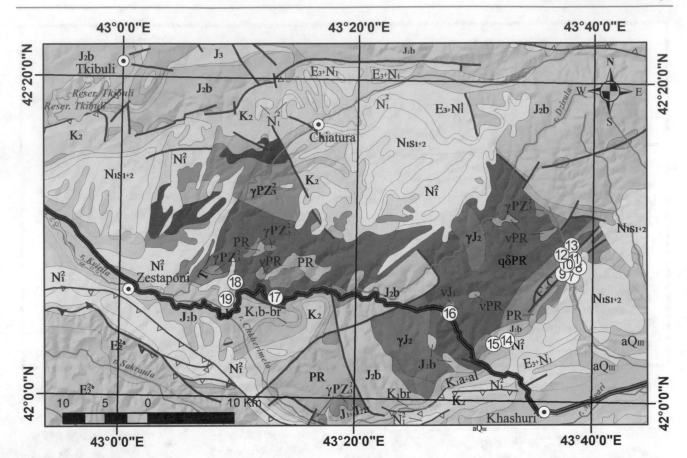

Fig. 3.22 Geotourist route Tbilisi-Zugdidi-Mestia-Ushguli (Khashuri-Zestaponi section). Clipping from the Geological map (1:500,000) of Georgia (Gudjabidze and Gamkrelidze 2003) showing the geological structure along the section and stop points (numbers in circles). Stratigraphic signs along the route: Q—Quaternary, N_1^2— Middle Miocene, K_2—Upper Cretaceous, K_1a-al—Aptian and Albian, K_1b-br—Beriasian-Barremian, J_2b—Bajocian, C-C—Cambrian, Silurian? Devonian and Carboniferous, qδPR—Proterozoic (orthogneisses), γPZ_3^2—Late Variscan granitoid intrusives

In the lens of marbles enclosed in the metamorphic schists of the Chorchana-Utslevi strip, archaeocytes were found pointed to the Early Cambrian age of the host rocks. Then, on the basis of the catagraphies observed in several marble lenses Early Cambrian age of the same part of the sequence was confirmed. According to the palynological data in this complex of metamorphites there are also Upper Silurian (?) and lower (?)—Middle and Upper Devonian palynomorphs revealed.

Special geological and petrologic-geochemical studies conducted in this part of the massif, allowed in a new way to decipher internal structure of metamorphic schist as well as express an opinion on the geological position and the age of ultrabasic and basic rocks closely associated with it (Gamkrelidze et al. 1981). In particular, petrologic and petrochemical studies have shown that the south-eastern part of the Dzirula massif among the Late Variscan granites preserved relics of the ophiolitic association (serpentinites, metabasites) of the Precambrian-Paleozoic age, belonging to the depleted upper mantle, as well as to the second and third

layers of the Paleotethys ocean, inherited developed from Prototethys (Gamkrelidze et al. 1981). Following this, in the work of Zakariadze et al. (1993) detailed research completely confirmed the typical ophiolitic nature of the metabasites noted above and the serpentinites. In particular, it was shown that metabasites correspond to series like MORB.

Ophiolites are represented by harzburgites and clinopyroxene-containing spinel ocean-type harzburgites— restites from melting of tholeiitic basalts, and metabasites, amphibolites and epidote amphibolites in composition of protolith correspond to tholeiitic basalts of N- and T-MORB types (Gamkrelidze and Shengelia 2005).

Based on the location on this section of the massif completely different in nature slices, in particular, fragments of the upper mantle and oceanic crust (serpentinites, amphibolites, gabbros, and gabbro-diabases), as well as of different age (Cambrian, Lower-Middle Paleozoic) and having different personal history of metamorphism of plates of sedimentary rocks (metavolcanogenic-phyllite complex) it

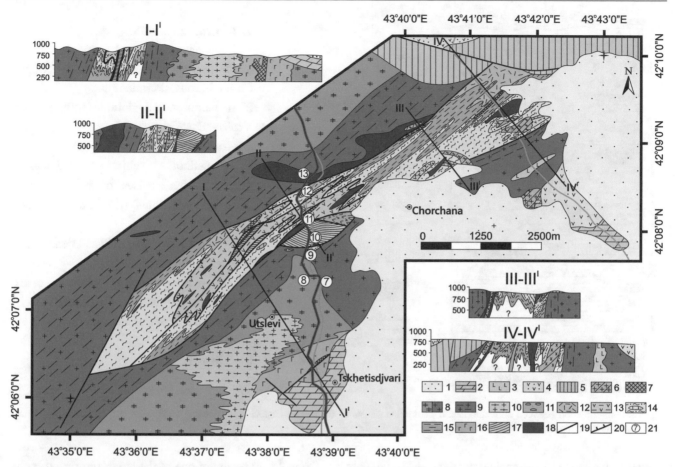

Fig. 3.23 Geological map of the area of Dzirula massif ophiolite outcrops (Gamkrelidze and Shengelia 2005). 1—Miocene terrigenous deposits; 2—Cretaceous carbonate deposits; 3—Bajocian diabase-porphyrite subvolcanic bodies; 4—Bajocian porphyrite series; 5—Liasic sandy-shale suite; 6—Upper Paleozoic subvolcanic bodies of rhyolites and granite porphyrites; 7—Upper Paleozoic dikes of alaskite granite; 8—a-Late Variscan K-feldspar granite (alaskite), b-Precambrian microclinized plagiogneissic and melanocratic basement; 9—milonitized granites; 10—Precambrian gneisses, migmatites, schists, amphibolites; 11 to 15—"schist suite": 11,12 —neo-autochthon: 11—Upper Carboniferous? to-Permian? Metamorphosed conglomerates, gravestones, sandstones, 12—Upper Visean-to-Bashkirian rhyolitic tuffs; 13 to 15—allochthon: 13—alternation of Middle?-to-Upper Devonian actinolite schists, metadiabases, metaporphyrites, basic tuff, and phyllites, 14—Lower Cambrian phyllites, sandy metashales, metasandstones, marble lenses, and quarzites, 15—Vendian?-to-Lower Cambrian mica-, two-mica-,garnet-, and chlorite-garnet schists; 16 to 18—Precambrian-to-Middle Paleozoic (?) allochthonous melanocratic complex: 16—gabbro and gabbro-diabase, 17—tectonic lenses and sheets of amphibolite, 18—serpentinite; 19—reverse and normal faults; 20 —overthrusts. Numbers in the circles—stop points

was concluded that all these rocks before the emplacement of the Variscan granites experienced a tectonic displacement (overthrusting) apparently during Bretonian orogeny because allochthonous plate contains Upper Devonian and overlapping it unconfirmably rhyolitic tuffs (neoauthochthon) belong in age to the Late Visean-Bashkirian stage. While cutting them granites are Late Variscan (Sudetian) in age. Thus, the "suite of metamorphic schists" is heterogeneous in composition and includes at least three different age groups of rocks. Rocks of the first two groups, dated paleontologically as the lower Cambrian and Middle Paleozoic, are apoterrigenous volcanogenic formations. Judging by the composition of clastic material, during the formation of primary sandy-clay deposits, the Precambrian complex, predominantly of plagiogneiss composition, was

the source of ablation of clastic material (Gamkrelidze and Shengelia 2005).

Despite the fact that the Chorchana-Utslevi allochthonous complex really consists of many small tectonic slice, we consider the most reasonable to distinguish of two main allochthonous plates of different ages: the Cambrian—Chorchana and the Middle Paleozoic—Ninissi, covered by a neo-autochthonous Cheshora suite.

During the Indosinian orogeny together with Sudetic granites and with the neo-autochthonous Chorchana suite, as well as allochtonic ophiolite complex experienced folding and intense tectonization, in particular milonitization of Sudetic granites in the near-contact strip with them (Fig. 3.23).

The Lower Middle Paleozoic part of the metamorphic schists, is a fragment of a volcanogenic-sedimentary lens

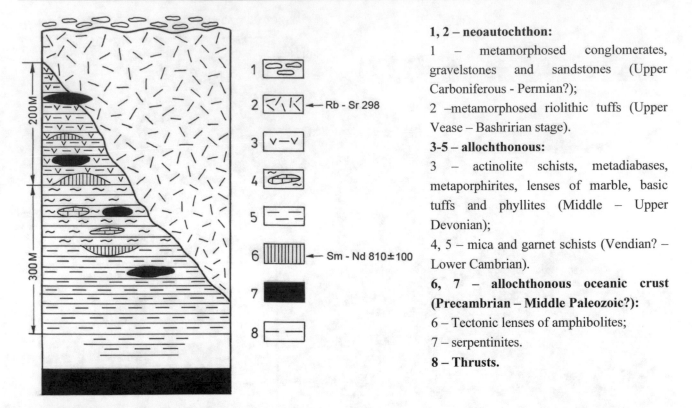

1, 2 – neoautochthon:

1 – metamorphosed conglomerates, gravelstones and sandstones (Upper Carboniferous - Permian?);

2 –metamorphosed riolithic tuffs (Upper Vease – Bashririan stage).

3-5 – allochthonous:

3 – actinolite schists, metadiabases, metaporphirites, lenses of marble, basic tuffs and phyllites (Middle – Upper Devonian);

4, 5 – mica and garnet schists (Vendian? – Lower Cambrian).

6, 7 – allochthonous oceanic crust (Precambrian – Middle Paleozoic?):

6 – Tectonic lenses of amphibolites;

7 – serpentinites.

8 – Thrusts.

Fig. 3.24 Interrelation between allochthonous metamorphic schists of Chorchana-Utslevi zone, oceanic ultrabasites and basites in the Dzirula crystalline massif (Gamkrelidze and Shengelia 2005)

formed in the limits of the ancient continental (or, possibly, island-arc) slope, which has subcontinental crust (Gamkrelidze and Shengelia 2005).

Familiarization with the structure of the Chorchana-Utslevi zone excursion provides along the r. Cheratkhevi (left tributary of the r. Suramula).

Stop 7. At the beginning the route crosses the rocks of granitized Precambrian plagiogneissic and melanocratic basement, which near the v. Tskhetijvari are covered transgressively with the Lower Cretaceous limestones. The Lower Cretaceous is represented here by Barremian lamestones of Urgonian facies.

Stop 8. Here crop out Late Variscan granites (alaskites).

Stop 9. Near the fault, bounding the Chorchana-Utslevi zone from the south, Late Variscan granites are intensively milonitized.

Stop 10. Exposures of tectonic slice composed of actinolite schists, metabasites, metaporphyrites, basite tuffs and phyllites of the Middle? and Upper Devonian.

Stop 11. Tectonic slice composed of phyllites, sandy metashales, metasandstones, marble lenses and quartzites of the Cambrian age cropping out in the right (western) bank of the river Cheratkhevi.

Stop 12. This slice is unconformable overlain by metamorphosed rhyolitic tuffs of Upper Visean-Bashkirian age.

Stop 13. Exposure of one of the larges protrusion bodies of serpentinites.

After this exposure, the excursion returns to the main Tbilisi-Kutaisi highway and continues to get acquainted with the pre-Alpine rocks of the Dzirula massif.

From the village Gomi the itinerary proceeds westwards to the Rikoti pass in order to acquaint excursion with the central part of the Dzirula inlier, which consist of the Late Variscan granitoids and Precambrian gneiss-migmatite complex, four- generation of metabasites and quartz-diorite orthogneisses.

The village Gomi—town Khashuri section of the route passes along the Kareli-Khashuri plain, representing the first terrace of the Kura river. In Surami, on the left bank of the river Suramula on the Middle Miocene sandstones an ancient fortress Suramis tsikhe of thirteenth century is erected.

In Surami the route proceeds along the left bank of the Suramula river and further up the Orkhevi river valley (left tributary of the Suramula river) to the village Chumateleti. Here in the narrow part of the gorge, on both banks crop out the Lower Cretaceous sediments, directly overlying granitoids of the Dzirula salient.

Stop 14. In the environs of the village Chumateleti the Upper Paleozoic complex of quartz-porphyries, granite-porphyries and alaskitic granites protrude from under the transgressive Urgonian limestones of the Barremian. The complex, which is the youngest one among Variscan granitoids of the Dzirula inlier, together with the volcanics of the Chiatura suite (quartz-porphyries) represent the plutonic-volcanic assemblage and is dated as the Late Paleozoic.

Stop 15. Within 1 km from the previous stop the oldest-Neoproterozoic gneiss-migmatite complex with U-Pb zircon 626 + 11 Ma (Gamkrelidze et al. 2011) crops out. It is intensely granitized by the Variscan granites. These are the microclinized chlorite-biotitic gneisses with remnants of amphibolites, injected by dikes and veins of granite-porphyries, quartz-porphyries and alaskitic granites.

Stop 16. The right bank of the river Rikotula—the granitized gneiss-migmatite complex injected by pla-giogranites and crossed by dikes and veins of the red gran-ites. It is represented by alternation of crystalline schists, plagiogranites and plagiomigmatites and amphibolites. Under the influence of Late Variscan granites they in places are granitized and transformed into granito-gneisses, granite migmatites and porphyroblastic microcline granites.

In 800 m down the Rikotula river the granitized complex is gradually replaced by plagiogneisses and then by "Riko-tites"—orthoclase gabbros (pyroxene, hornblende, biotite of labrador, orthoclase) of Triassic age (Shengelia et al. 2012). Orthoclase seems to impregnate the rock. Veins and dikes of the red granites are observed.

Stop 17. In the environs of the village Ubisa the granite-migmatite complex comprises dikes and veins of the red granites, granite-pophyries, quartz-porphyries and aplitic granites.

Stop 18. In the village Shrosha one of the pegmatite fields of the Dzirula inlier occures. About 500 pegmatitic and aplitic veins are known here. The largest pegmatitic veins of the Shrosha field attain 25–30 m in thickness.

West of the village Shrosha the Dzirula crystalline basement begins to sink gradually beneath the Mesozoic and Cenozoic sedimentary cover (quartz-sandstones and rhy-olitic volcanics of the Lower Liassic, Middle-Upper Liassic red crinoidal limestones, Bajocian tuffaceous rocks, etc.).

Stop 19. Near the v. of Shorapani, Lower Jurassic-Aalenian and then Neogene trerrigenous rocks belonging to the sedi-mentary cover of the Dzirula crystalline massif crop out.

Stop 20. Further, near the city of Kutaisi (Fig. 3.25), the route crosses alkaline volcanics of so-called Mtavari suite of Upper Cretaceous age. It is composed of the picrite-basalts, olivine and analcite basalts, trachy-basalts and tuffs with sedimentary interbeds.

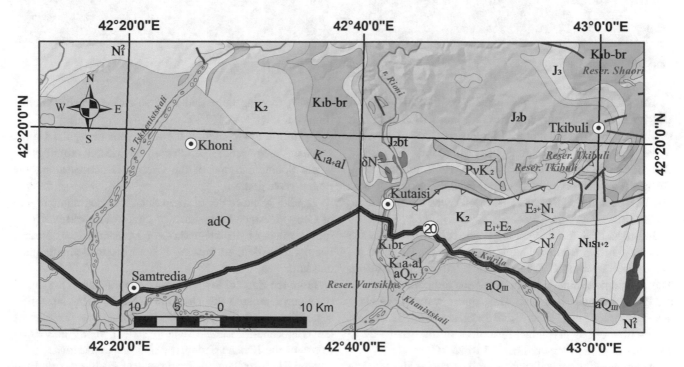

Fig. 3.25 Geotourist route Tbilisi-Zugdidi-Mestia-Ushguli (Zestaponi-Samtredia section). Clipping from the Geological map (1:500,000) of Georgia (Gudjabidze and Gamkrelidze 2003) showing the geological structure along the section and stop points (numbers in circles). Stratigraphic signs along the route: Q—Quaternary, N_1s_{1+2}— Lower and Middle Sarmatian, N_1^2—Middle Miocene, $E_2^3 + E_3$—Upper Eocene-Oligocene, E_2^3—Upper Eocene, E_2—Middle Eocene, $E_1 + E_2$—Paleocene and Eocene, $E_1 + E_2^1$—Paleocene and Lower Eocene, K_2—Upper Cretaceous (Mtavari suite), K_1a-al—Aptian and Albian

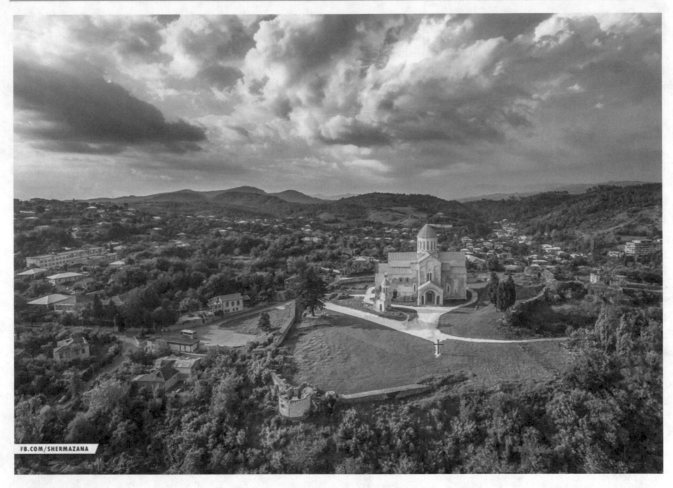

Fig. 3.26 Bagrati Cathedral (photography by Giorgi Shermazanashvili)

Kutaisi functioned as the capital of the Kingdom of Colchis in the sixth–fifth centuries BC. From 1008 to 1122 it serves as capital of the united Kingdom of Georgia and then it was the capital of the Imeretian Kingdom. The ancient Bagrati fortress and church built in 1003 by the King Bagrat III are preserved in the city (Fig. 3.26). From October 2012 to December 2018, Kutaisi briefly was the seat of the Parliament of Georgia.

From Kutaisi the excursion proceeds westwards to the Tskaltubo health resort, which is situated in the valley of the karst river Tskaltubo. Tskaltubo is a balneological resort. The radioactive mineral water gryphons are concentrated in the old Tskaltubo river bed. The Albian water-proof clays are developed here. Where the clays are eroded, thermal waters flow out as gryphons. Resources of the water—22 to 23 lt per day, the temperature—34 to 35 °C.

After that at a great distance the rout passes through the Quaternary deposits of the Western Rioni (molasse depression) (Fig. 3.27).

Beginning from the v. of Senaki the rout crosses Paleocene-Eocene, Oligocene and Miocene terrigenous deposits, forming large and wide so-called Odishi syncline (Fig. 3.27).

Up to the town Zugdidi the route runs along the Odishi syncline, then it turns northward to Mestia crossing the whole Southern Slope of the Greater Caucasus along the Enguri river gorge.

Zugdidi is the center of Megrelia, an ethnographic region of Georgia. Among the sights of the town an old palace of former rulers of the Megrelia—the Dadiani princes attracts visitors' attention. At present, it is a historical-ethnographic museum.

From the Zugdidi town the route running along the Inguri river gorge crossed the Neogene deposits. They are represented by coarse molasses, sandy-argillaceous deposits of the Oligocene-Miocene (Maikop series), marls and limestones of the Eocene, Paleocene and the Cretaceous.

Stop 21. Near the v. of Jvari one can see big flexure built up of Upper Jurassic terrigenous, Cretaceous carbonaceous and Paleogene terrigenous rocks. The flexure reflects on the surface a hidden deep faulte at a depth, which represents the boundary between Georgian block and the Gagra-Java zone

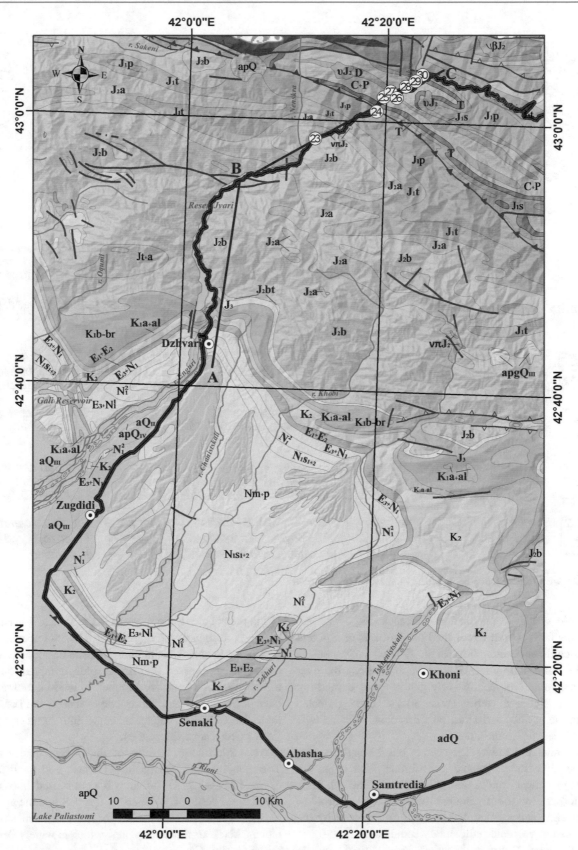

Fig. 3.27 Geotourist route Tbilisi-Zugdidi-Mestia-Ushguli (Samtredia-Lakhamula section). Clipping from the Geological map (1:500,000) of Georgia (Gudjabidze and Gamkrelidze 2003) showing the geological structure along the section and stop points (numbers in circles). A–B–C—segmented line of geological profile. Stratigraphic signs along the route: Q—Quaternary, Nm + p—Meotian and Pontian, N_1s_{1+2}—Lower and Middle Sarmatian, N_1s—Sarmatian, N_1^2—Middle Miocene, $E_2^3 + E_3$—Upper Eocene-Oligocene, E_2^3—Upper Eocene, E_2—Middle Eocene, $E_1 + E_2$—Paleocene and Eocene, $E_1 + E_2^1$—Paleocene and Lower Eocene, K_2—Upper Cretaceous, K_1a-al—Aptian and Albian, K_1b-br—Beriasian-Barremian, J_3—Upper Jurassic, J_2bt—Batonian, J_2b—Bajocian, J_2a—Aalenian, J_1t—Toarcian, J_1P—Plinsbachian, J_1s—Sinemurian, Tm—Triassic, C + P—Carboniferous and Permian, D—Devonian

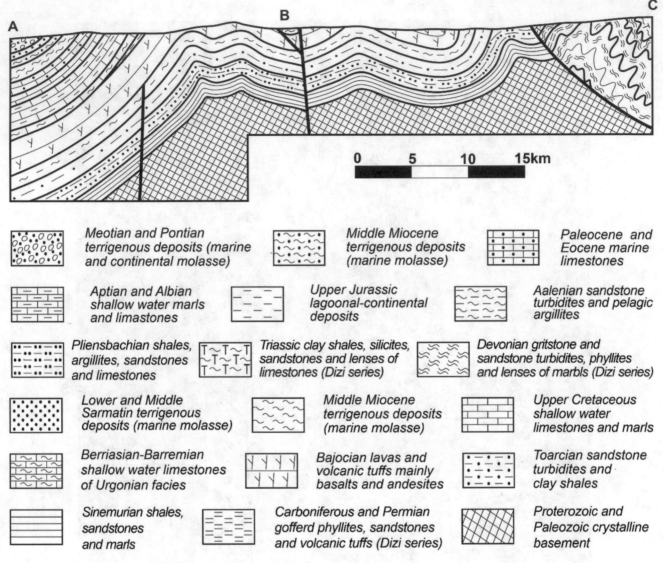

Legend:

- Meotian and Pontian terrigenous deposits (marine and continental molasse)
- Middle Miocene terrigenous deposits (marine molasse)
- Paleocene and Eocene marine limestones
- Aptian and Albian shallow water marls and limastones
- Upper Jurassic lagoonal-continental deposits
- Aalenian sandstone turbidites and pelagic argillites
- Pliensbachian shales, argillites, sandstones and limestones
- Triassic clay shales, silicites, sandstones and lenses of limestones (Dizi series)
- Devonian gritstone and sandstone turbidites, phyllites and lenses of marbls (Dizi series)
- Lower and Middle Sarmatin terrigenous deposits (marine molasse)
- Middle Miocene terrigenous deposits (marine molasse)
- Upper Cretaceous shallow water limestones and marls
- Berriasian-Barremian shallow water limestones of Urgonian facies
- Bajocian lavas and volcanic tuffs mainly basalts and andesites
- Toarcian sandstone turbidites and clay shales
- Sinemurian shales, sandstones and marls
- Carboniferous and Permian gofferd phyllites, sandstones and volcanic tuffs (Dizi series)
- Proterozoic and Paleozoic crystalline basement

Fig. 3.28   A-B-C cross-section between Jvari-Lakhamula (see Fig. 3.27)

of the Fold system of the Greater Caucasus (Figs. 3.27 and 3.28). This fault is well established on the basis of different geophysical data (Gamkrelidze et al. 1998).

To the north, within the Gagra-Java zone along the river Enguri the route crosses relatively gentle folds, composing of Lower Jurassic terrigenous rocks and Bajocian calk-alkaline basalts, andesites, lava breccias, volcanic tuffs and, in places, tephroturbidites.

From under the Lower Cretaceous basal conglomerates the Upper Jurassic variegated gypsiferous suite crops out consisting of clays, sandstones, gravelstones (260 m). They are underlain by the Bathonian coal-bearing aleurolites, argillites and sandstones down the section replaced by the thick Bajocian volcanic-sedimentary sequence.

The Bajocian "porphyritic suite" builds up the whole midstream of the Enguri river, cropping out over some tens of kilometers along the road up to the village Khaishi. Here it is represented predominantly by tuffaceous rocks, belonging to the calc-alkaline andesite-basaltic series. The rocks are compressed into linear folds, compression degree increasing northward. A number of sub-latitudinal faults have been established. Along one of these—the Larakvakva fault the Bajocian volcanics come into contact with the Lower Cretaceous limestones.

Stop 22. In the vicinity of the village of Khaishi the Toarcian-Aalenian Sori suite of slates and sandstones is exposed (Fig. 3.27). The suite is compressed into sub iso-clinal linear folds of W-E strike. It represents the upper part of the Lower Jurassic-Aalenian series of sandstone turbidites, black argillaceous slates, which is widely developed in the Greater Caucasus.

Stop 23. Leaving behind Khaishi the outcrops of the Bajocian porphyritic suite can be seen in the core of the isoclinal steep syncline. Unlike the previous section in this terrain the suite consists predominantly of lavas, often of pillow-lavas. In the northern limb of the syncline the Aalenian Sori suite comprises thick sills of diabase-porphyrites (near the village of Jorkvali). Apparently, these diabase-porphyrites are synchronous to the Bajocian porphyritic suite.

From the Jorkvali village up to the first outcrops of the Dizi series the route crosses the lower horizons of the slate series of the Lower Jurassic: the black argillaceous slate suite (Pliensbachian), the clay shale and sandstone suite (Sinemurian) compressed into sub-isoclinal linear southward overturned folds.

Stop 24. Outcrops of the Dizi series begin at this point (Figs. 3.27 and 3.28).

The Dizi series composed of faunistically dated weakly metamorphosed mostly terrigenous sediments with a thickness of 1800–2000 m. According to most researchers they include continuous section from the Devonian to the Triassic. In the composition of the Dizi series on the basis of Conodonts revealed three suites (Kutelia et al. 1984): Kirari (Devonian), Tskhenistskali (Carboniferous-Permian) and Gvadarashi (Triassic), composing a series of highly compressed folds complicated by minor folding and faults. According to these authors, the lowest suite Devonian is composed mainly of dark-gray gritstones, coarse-grained sandstones and phyllites. At the top of the suite shales, lenses of marbles, intraformational conglomerates, interlayers of siliceous shales and silicates are developed. Middle suite (carboniferous-Permian) is represented by corrugated phyllites, silver-gray and brownish sandstones, benches of volcanic rocks, lenses of marbles, interlayers of siliceous shales and silicites. The upper (Triassic) suite is composed of black shale, silicites, sandstones and gritstones, limestone lenses.

According to the definitions of conodonts in the section of this complex, all stages from the Eifelian to the Visean are established (Kutelia 1983)

At the very top of the section based on the stratigraphic position and spore-pollen analysis the Triassic is established (Kutelia et al. 1984).

It is noteworthy that in the Dizi series, lateral facies variability is noted. In particular decrease from south to north of the size of clastic material, which indicates the presence of areas of disintegration to the south of its modern exposures.

Within the series folds with northern vergence are developed (Fig. 3.27). They are metamorphosed into the green schist facies in the pre-Early Jurassic time because according to Somin (1971), and Gamkrelidze (1991) in the north and the southern edges of the Dizi series Jurassic conglomerates, with already metamorphosed and deformed pebbles, overlap various horizons of Dizi series—from the Devonian to the Triassic.

The Dizi series is generally characterized by an imbricate structure. Cleavage and faults in the Paleozoic strata of this series, despite the general overturning of Jurassic strata to the south on the entire southern slope of the Greater Caucasus, are steeply inclined to the south, and the folds preserved the northern vergence. This indicates, all that they are displaced (overthrust) to the north in the Early Cimmerian (Indosinian) orogeny, which caused them to be in direct contact in the north with highly metamorphosed rocks of the same age of the Main Range zone of the Greater Caucasus (see Fig. 2.2).

Deposits of Dizi series were formed in the continental slope and foot on the southern passive margin of the small ocean basin of the southern slope of the Greater Caucasus (Gamkrelidze and Shengelia 2005).

In the Enguri river valley the Dizi series is composed in a core of the E-W trending highly compressed sub isoclinal composite anticline (Fig. 3.28). It is overlain by the Lower Liassic deposits, in which the Sinemurian is proved by ammonites.

Stop 25. North of the confluence of the Kazakhtvib and Enguri rivers the tectonic contact of the Lower Jurassic with the Dizi series is exposed. The Sinemurian argillaceous slates are crushed and hydrothermally altered. Just beyond the fault the Dizi series is composed of the dark-grey silicites (5–10 m), containing poorly preserved radiolarians of Mesozoic habitus. Further there follow: phyllitized slates with boudins of sandstones (40–50 m), black argillaceous slates, gritstones (5–6 m), black slates with sandstone boudins and hydrothermally altered areas (Gvadarashi suite).

Stop 26. Greenish, silver-grey, light-brown sandstones, gritstones and phyllites. Plication, small slice folding and corrugation of phyllites occur. Diabase-porphyritic dikes are common (Tskhenistskali suite, Carboniferous-Permian). In the parallel sections coral and conodonts are recognized in silicites.

Stop 27. Sandstones, gritstones, phyllites, interbeds of silicites are outcrop here. On opposite bank of the gorge they contain a lens of the dark-grey limestones with Middle and Upper Devonian fauna (corals, crinoids) (Kutelia et al. 1984).

On the right bank the Frasnian and Famennian conodonts are known in the interbeds of silicites (the section is ascending).

Stop 28. The Dizi village. A sequence of quartz-biotite hornfels, intruded by diorite-monzonite bodies, includes large lenses of white banded marble being worked in a small quarry.

Stop 29. The Kirari suite: sandstones, gritstones, phyllites, silicites. The latter contain the Middle Devonian (Eifelian) conodont fauna. Further on grey phyllites are exposed with marble lenses, containing the Lower Permian fauna of small foraminifera (Tskhenistskali suite). In southern limb of an anticlinorium synchronous deposits are represented by green corrugated phyllites and light-brown and light green sandstones.

Stop 30. The northern contact of the Dizi series with basal sequence of the Lower Jurassic. The latter is represented here by thick-bedded quartz-arkose sandstones, gritstones, quartz-keratophyric volcanic and argillaceous slates. The rocks are intensely altered hydrothermally.

The lowermost dated part of the Liassic that contains Sinemurian ammonites up the gorge is followed by the Pliensbachian slates. Further upstream up to the village of Mestia the route crosses the higher horizons of the Mesozoic. In the vicinity of Mestia it enters the area of spreading of the Upper Jurassic—Lower Cretaceous carbonate flysch, represented by cleaved marls and thin-laminated limestones. They crop out in the core of a highly compressed E–W trending syncline, complicated by smaller sub isoclinal southward overturned folds and southward directed upthrusts resulting in imbricated slice structure.

In the section Lakhamula-Ushguli (Fig. 3.29) the road follows mainly along the strike of sandstone and clastic-limestone flysch deposits of Berriasian and Valanginian stages and Lower substage of Hauterivian.

Stop 31. Mestia—the district center of the Upper Svaneti, one of the Alpine ethnographic regions of Georgia. Despite its small size, the town was an important center of Georgian culture for centuries and contains a number of medieval monuments, such as churches and forts, included in a list of UNESCO World Heritage Sites. The town is dominated by stone defensive towers of a type seen in Ushguli ("Svan towers"). Ushguli is located at the foot of Mount Shkhara (5201 m), at the confluence of the Enguri and Shavtskala-Kvishara. The height of Ushguli community varies from 2060 to 2200 m above sea level. According to these data, Ushguli was considered to be the highest settlement in Europe until 2014 (Gamkrelidze et al. 2019).

To the north-east from the v. of Mestia along the rivers Mulkhra (eastern tributary of the r. Enguri) and Mestiachala crop out Berriacian-Hauterivian and Upper Jurassic clastic-limestone and rarely sandstone-siltstone turbidites, marls and clay shales belonging to Mestia-Tianeti (Flisch) zone of the Greater Caucasus (Fig. 3.29).

Ushguli village, which is located in the r. Enguri sources, being one of the highest continuously inhabited settlements in Europe. The ninth–twelfth centuries famous 30 towers of village have been included in the UNESCO's World Heritage list.

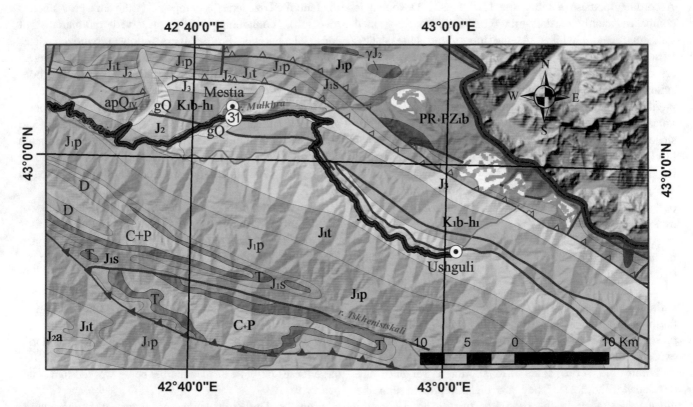

Fig. 3.29 Geotourist routeTbilisi-Zugdidi-Mestia-Ushguli (Lakhamula-Ushguli section). Clipping from the Geological map (1: 500,000) of Georgia (Gudjabidze and Gamkrelidze 2003) showing the geological structure along the section and stop points (numbers in circles). Stratigraphic signs: K_1b-h_1—Berriasian, Valanginian and lower substage of Hauterivian. Other signs see in Fig. 3.27

Fig. 3.30 The upper reaches of the river Enguri, village Ushguli and the granitoid massif Shkhara (photography by Daniel Tormey). The highest point is 5201 m. The age is 320 Ma (dated by U–Pb single zircons method). Dashed line indicates the main thrust of the Greater Caucasus (Gamkrelidze et al. 2019)

Ushguli is overlooked by magnificent 5100–5200 m high and 15 km long Upper Paleozoic Shkhara granitoid massif covered with perpetual snow and glaciers. Near the Inguri river source, between the Shkhara Paleozoic massif and the Lower Jurassic shales crops out the Main Thrust zone of the Greater Caucasus which is easily accessible on foot (Fig. 3.30). Here we can observe how this grandiose crystalline massif thrusts over sedimentary rocks.

3.3 Tbilisi-Khashuri-Borjomi-Vardzia

At the beginning participants of geotoure get acquainted with the sights and geological structure of the environs of the city of Tbilisi, and then to the town of Mtskheta with the geology of the eastern termination of the Adjaro-Trialeti folded zone. First 6 stop points are described above (see chapter Tbilisi-Pasanauri-Kazbegi (along the Georgian Military Road), ("Environs of Tbilisi").

Further, along the Mtkvari (Kura) river, crop out Neogene and Quaternary molasse deposits of Kartli intermountain depression.

After the town of Khashuri, near the village of Akhaldaba excursion starts crossing the Adjara-Trialeti folded zone (Figs. 3.31 and 3.32).

The Adjara-Trialeti folded zone is a clearly delineated structural-morphological unit, elongated in the latitudinal direction between the eastern coast of the Black Sea and the Iori Upland.

The oldest deposits of zone are represented by thick (more than 2000 m) Albian-Lower Turonian and in the southern part of zone—Cenomanian-Lower Maastrichtian volcanogenic formation predominantly basaltic composition. Above follow carbonaceous rocks of the Cretaceous (300–1000 m), which are transgressively overlapped by the Paleocene-Lower Eocene mainly sandstone-siltstone flysch formation (1500–2000 m). Above follows the Middle Eocene volcanogenic formation, thickness of which varies from 500–600 m in the east increases to the west to 5000 m. In the western part of zone along the axial strip is represented by tholeiitic basalts and along the edges—by subalkaline basalts in the south and potassium alkaline effusions in the north. The Upper Eocene and Oligocene—Lower Miocene consist of terrigenous deposits and only in the south-eastern part of zone they are replaced by volcanites of medial composition (Gamkrelidze 1976).

In modern structure Adjara-Trialeti folded zone represents anticlinorium with rather diverse morphology of composing its structures. The longitudinal deep faults planned within the zone divide it on elevated central (axial block) and relatively submerged north and south subzones. Besides, according to character of fold structures, within the zone can be distinguished western, central and eastern segments (Gamkrelidze 1976).

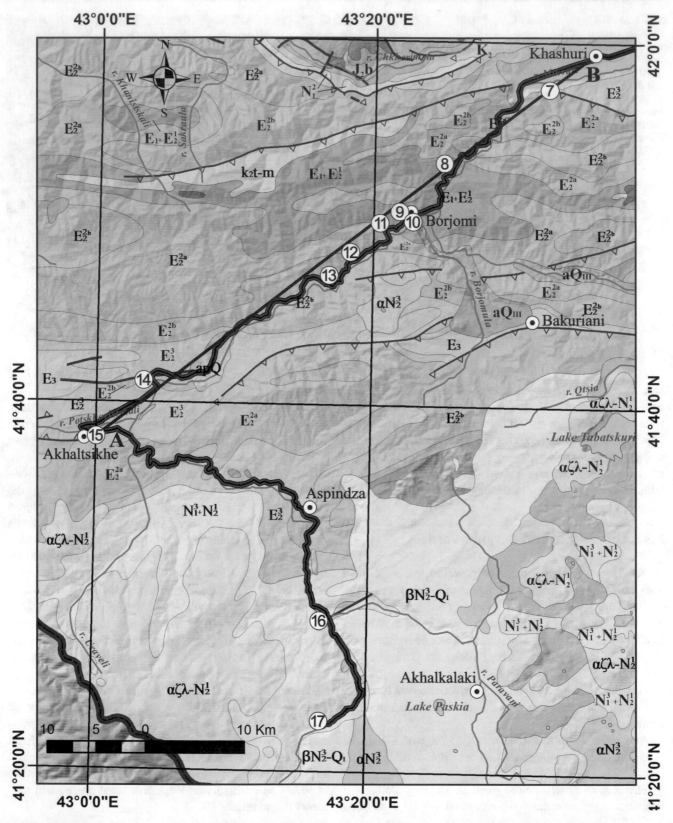

Fig. 3.31 Geotourist route Tbilisi-Khashuri-Borjomi-Vardzia (Khashuri-Vardzia section). Clipping from the Geological map (1:500,000) of Georgia (Gudjabidze and Gamkrelidze 2003) showing the geological structure along the section and stop points (numbers in circles). AB—line of geological profile. Stratigraphic signs: aQ_{III}—Upper Quaternary andesitic lava flows, βN_2^3-Q_1—Upper Pliocene-Quaternary continental basaltic, doleritic and andesite-basaltic lava flows. $N_1^3 + N_2^1$—Upper Miocene and Lower Pliocene continental deposits (Volcanic breccias, tuff-diatomites, andesitic and basaltic sheets(lower part of the Goderdzi suite), E_3—Oligocene, E_2^3—Upper Eocene, E_2^{2b}—upper part of the Middle Eocene, E_2^{2a}—lower part of the Middle Eocene, $E_1 + E_2^1$—Paleocene and Lower Eocene

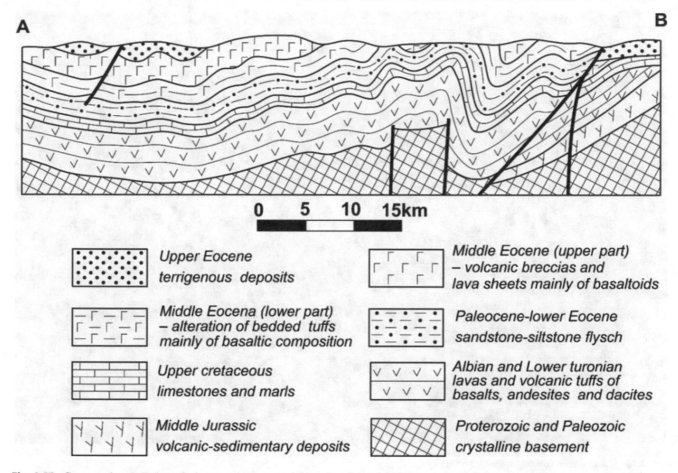

Fig. 3.32 Cross-section A–B through the central segment of the Adjara-Trialeti folded zone showing the presumable structure at a depth

Western segment is characterized by generally uncomplicated structure—within it several large gentle folds are developed, which are complicated with strike-slip faults.

In contrast to the western segment, the central segment has a more complex structure with a clearly manifested axial block uplift and developed on both sides of it rather long and tight folds, which substitute each other along the strike. They are complicated with thrusts inclined to the north. In this part of zone, it is overthrust to the north. But to the east, along the southwestern periphery of the Dzirula massif, the fault has a northwestern strike and right-lateral character.

In the region of the eastern sinking of Adjara-Trialeti zone (eastern segment), a general simplification of its structure occurs. Here, the folds gradually sink to the east under the Miocene-Pliocene cover of the Kura depression.

Stop 7. Familiarization with the structure of this zone begins near the village of Akhaldaba, where one can see the tectonic overlapping of Upper Eocene terrigenous deposits by the Middle Eocene particolored hetero-clastic tuffs of basaltic composition (Fig. 3.32) (Gamkrelidze 1976).

To the south, Middle Eocene rocks form a rather large symmetrical so-called Akhaldaba syncline and Bebrisi anticline.

Stop 8. Near the v. of Rveli Paleocene-Lower Eocene flysch (Borjomi flysch) crops out. It is represented mainly by rhythmical alternation of sandstone-siltstone turbidites, tephroturbidites and clays. They form here clearly asymmetrical Lomismta anticline with very steep northern and very gentle southern limbs (Fig. 3.32).

This fold represents a reflection on the surface of the northern limitation of the axial block uplift of the Adjara-Trialeti zone (Gamkrelidze 1976). Further, one can observe almost uninterrupted outcrops of Borjomi flysch rocks, which form here many minor folds and faults (Gamkrelidze 1976).

Stop 9. Surface reflection of the southern limitation of the axial block of Adjara-Trialeti zone at a depth represents Borjomi anticline, which is well exposed in environs of well-known resort town Borjomi (detailed description of Borjomi resort as potential geopark see below).

Fig. 3.33 Rabati Castle (photography by I kynitsky)

Stop 10. In Borjomi flysch deposits are overlapped by Quaternary andesitic lava flows.

Stop 11. Near the v. of Likani, Borjomi flysch is conformably overlain by Middle Eocene thin-bedded tuffs of the so-called Likani suite consisting of parti-coloured heteroclastic tuffs of basaltic composition. They belong to the lower part of the Middle Eocene.

Stop 12. Upper part of thin-bedded Likani suite is represented in this segment of zone by very thick lava sheets of dellenites (Kvabiskhevi suite) (Gamkrelidze 1976).

Stop 13. Upper part of the Middle Eocene—the so-called Dviri suite crops out in rather steep Dviri syncline and is composed of rudaceous, thick-bedded, massive volcanic rocks and lava sheets of subalkaline, and calc-alkaline basaltoids (Gamkrelidze 1976).

Stop 14. Near the v. of Atskuri Dviri suite is overlapped by terrigenous deposits of the Upper Eocene. Within the Upper Eocene One can see a small exposure of the Middle Eocene Dviri suite, which form a symmetrical anticline.

Stop 15. Akhaltsikhe (literally "new castle") is a small city in Georgia's southwestern region of Samtskhe-Javakheti. It is situated on the both banks of a small river Potskhovi, which separates the old city in the north from the new city in the south.

In the old part of the city one can see the great Rabati Castle, built by the Ottomans around a mosque, and St. Marine Church (Fig. 3.33). The hills nearby the city harbour the Sapara Monastery (tenth–fourteenth centuries).

As a result of the ongoing restoration in 2011–2012, the fence, citadel, IX–X century Orthodox Church, Haji Ahmed-Pasha Jakeli Mosque, madrasa, minaret and Jakeli Palace were renovated on the territory of the castle-complex, where the Samtskhe-Javakheti History Museum was located (Fig. 3.33).

Leaving Akhaltsikhe to the south-east along the Akhaltsikhe-Vardzia road, the excursion crosses the rather wide exposures of the Upper part of the Middle Eocene and then in the Aspindza syncline—Upper Eocene terrigenous deposits.

Stop 16. Then, near the v. of Toloshi crop out Upper Miocene and Lower Pliocene continental deposits: tuffs, volcanic breccias, conglomerates, diatomites, calc-alkaline andesitic and basaltic sheets, belonging to the so-called Goderdzi suite. Just in these rocks the Vardzia rock city is hewn.

Stop 17. Relicts of the Late Miocene mega volcano and its pyroclastic flows, in which amazing city Vardzia is hewn in the solid rock are descript in detail in Okrostsvaridze et al. (2016, 2019).

3.4 Summary

Main geotourist routes of Georgia are: (1) Tbilisi-Pasanauri-Kazbegi (along the Georgian Miliraty Road), (2) Tbilisi-Zugdidi-Mestia-Ushguli and (3) Tbilisi-Khashuri-Borjomi-Vardzia. First route crosses Paleogene sedimentary and volcanic deposts of the eastern termination of the Adjara-Trialeti folded zone, weekly folded Neogene, Paleogene deposits of the Kura intermountain depression and intensive folded Mesozoic and Cenozoic deposits and tectonic nappes of the Fold System of the Greater Caucasus. Second route crosses Neogene, Paleogene deposits of the Kura intermountain depression, pre-Alpine metamorphic rocks of the Dzirula massif and intensive folded Cenozoic, Mesozoic and weekly metamorphosed Paleozoic rocks of the Greater Caucasus. Third route crosses Paleogene sedimentary and submarine volcanic rocks of central part of the Adjara-Trialeti folded zone, Upper Miocene and Lower Pliocene continental deposits and continental volcanic rocks of the Artvin-Bolnisi zone.

References

Gamkrelidze IP (1976) Mechanism of tectonic structure formation and some general problems of tectogenesis. Proc Geol Inst Acad Sci Georgia, New Series 52:1–226 (in Russian with extended English summary)

Gamkrelidze IP (1984) The Outscirts of Tbilisi. In: Guide-book of geological excursion of the International Geological Congress, 27-th Session, Georgia, pp 66–77

Gamkrelidze IP (1991) Tectonic nappes and horizontal layering of the Earth's crust in the Mediterranean belt (Carpathian, Balkanides and Caucasus). Tectonophysics 196(3–4):385–396

Gamkrelidze PD, Gamkrelidze IP (1977) Tectonic nappes of southern slope of the Greater Caucasus. Publishing House "Mecnireba", Tbilisi, pp 1–87 (in Russian)

Gamkrelidze IP, Kandelaki DN (1984) Guide-book of geological excursion of the International Geological Congress, 27-th Session, Georgia, pp 41–55

Gamkrelidze I, Maisadze F (2010) Some new considerations on the age, composition, geological position and genesis of olistostromes of the southern slope of the Greater Caucasus. Bull Georgian Natl Acad Sci 4(2):103–116

Gamkrelidze IP, Maisadze FD (2016) Formation conditions of upper Eocene olistostromes and retro-overthrusts at the southern slope of the Greater Caucasus. Geotectonics 50:598–607. https://doi.org/10.1134/S0016852116060029

Gamkrelidze IP, Shengelia DM (2005) The Precambrian-Palaeozoic regional metamorphism, magmatism and geodynamics of the Caucasus. Publishing House "Scientific World", Moscow, pp 1–458. (in Russian with extended summary)

Gamkrelidze IP, Dumbadze GD, Kekelia MA, Khmaladze II, Khutsishvili OD (1981) Ophiolites of the Dzirula massif and the problem of the Paleotethys in the Caucasus. Geotektonics 5:23–33

Gamkrelidze I, Giorgobiani T, Kuloshvili S, Lobjanidze G, Shengelaia G (1998) Ative deep faults map and catalogue of the territory of Georgia. Bull Georgian Natl Acad Sci 157(1):80–85

Gamkrelidze I, Nadareishvili G, Tsamalashvili T, Basheleiscvili L (2008) On genesis of Tbilisi olistostromes. Proc Geolo Inst Acad Sci Georgia, New Ser 124:24–29

Gamkrelidze I, Maisadze F, Nadareishvili G (2010) Guide book of geological excursion of the International Scientific Conference "Problems of geology of the Caucasus". Tbilisi, pp 1–35

Gamkrelidze I, Shengelia D, Tsutsunava T, Chung SL, Chiu HY, Chikhelidze K (2011) New data on the U-Pb zircon age of the pre-Alpine crystalline basement of the Black-Sea- Central Transcaucasian terrane and their geological significance. Bulletin of the Georgian National Academy of Sciences 5(1):64–76.

Gamkrelidze I, Okrostsvaridze A, Maisadze F, Basheleisvili L, Boichenko G, Skhirtladze I (2019) Main features of geological structure and geotourism potential of Georgia, the Caucasus. J Mod Environ Sci Eng 5(5):422–442. https://doi.org/10.15341/mese (2333-2581)/05.05.2019/010

Gudjabidze G, Gamkrelidze I (2003) Geological map of Georgia (scale 1:500,000) (Editor I. Gamkrelidze), Tbilisi

Kutelia ZA (1983) New data on stratigraphy of Dizi series (Svaneti). Bull Acad Sci GSSR 109(3):29–33 (in Russian)

Kutelia ZA, Adamia Sh, Ananiashvili GD (1984) Tkaltubo-Zugdidi-Mestia. In Guide-book of geological excursion. In: 27th Session on International Geological Congress, Moscow, pp 146–151

Maisadze FD (1994) On the upper Eocene Olistostromes of the Southern Slope of the Greater Caucasus. Stratigr Geol Correl 2 (1):95–102 (in Russian)

Maisadze FD (2008) Event deposits in chaotically built formations. Bull Georgian Natl Acad Sci 2(3):79–87

Maisadze FD (2016) Some data on the rate of sedimentation. Bull Georgian Natl Acad Sci 7(2):79–87

Mauvilly J, Koiava K, Gamkrelidze I, Mosar J (2015) Tectonics in the greater Caucasus: a N-S section along the Georgian Military Road—Georgia. In: Proceedings of the 13th Swiss Geoscience Meeting, Basel, Switzerland. https://doi.org/10.13140/RG.2.2.33251.45607

Mauvilly J, Koiava K, Gamkrelidze I, Mosar J (2016) Tectonics in the Georgian Greater Caucasus: a structural cross-section in an inverted rifted basin setting. In: Proceedings of the 14th Swiss Geoscience Meeting, Geneva, Switzerland. https://doi.org/10.13140/rg.2.2.26540.56963

Meladze GK (1967) Hipparion fauna of Arkneti and Bazaleti. Publishing House "Mecniereba", Tbilisi, pp 1–168. (in Russian)

Okrostsvaridze A, Elasvili M, Popkhadze N, Kirkitadze G (2016) New Data on the Geological Structure of the Vardzia Cave City, Georgia. Bulletin of the Georgian National Academy of Sciences 10(3):98–105

Okrostsavridze A, Chung SL, Lin YC, Skhirtladze I (2019) Geology and Zircon U-Pb Geochronology of the Mtkvari Pyroclastic flow and evaluation of destructive processes affecting Vardzia rock-cut city, Georgia. Quaternary International 540:137–145. https://doi.org/10.1016/j.quaint.2019.03.026

Shengelia D, Shubitidze L, Chung Sun-Lin, Chiu Han-Yi, Treloar P (2012) New data on the formation and age of orthoclase gabbro of the Dzirula massif (Georgia). Bull Georgian Natl Acad Sci 6(3):75–82

Somin M (1971) Pre-Jurassic basement of the Main range zone and southern slope of the Greater Caucasus. Publishing House "Nauka", Moscow, pp 1–245. (in Russian)

Zakariadze G, Adamia Sh, Kolcheva K (1993) Geochemistry of metabasites series of pre-Alpine ophiolites of the Eastern Mediterranean Region. Petrology 1(1):50–87 (in Russian)

As the potential Geoparks we could consider: (1) Sataplia dinosaur footprints together with Sataplia and Prometheus caves; (2) Tskaltubo resort-town and mineral water deposit; (3) Borjomi resort-town and mineral water deposit; (4) Kazbegi Quaternary volcano and Keli volcanic highland; (5) Dariali Paleozoic granitoid massif; (6) Dmanisi hominids site and the Mashavera gorge basaltic flow; (7) Dashbashi canyon; (8) Uplistsikhe rock-cut town and Kvakhvreli cave complex; (9) Udabno—Upper Miocene marine and continental deposits and David Gareja monastery complex; (10) Dedoplistskaro—Vashlovani protected areas and mud volcanoes.

Another potential Geopark—Vardzia Upper Miocene megacaldera and ancient Vardzia rock-cut town is described in detail in the spetial works of Okrostsvaridze et al. (2016, 2019) and therefore is not considered here.

4.1 Sataplia Dinosaur Footprints Together with Sataplia and Prometheus Caves

Sataplia is mountainous area, which is located in Western Georgia, near the city of Kutaisi (5 km), 250 km from Tbilisi. In Georgian Sataplia means "place of honey" and this name gained from the wild honey, which is a lot in the local forest. There is a rare combination of important natural sites in this area: karst caves, dinosaur footprints and botanical attractions. That's why in 1935 here was created State Natural Reserve of Sataplia. The reserve is located on the edge of the southern slope of the Greater Caucasus, in 8 km from Tskaltubo resort. Total area of the Nature Reserve is 354 ha (the highest top—520 m), 209 ha from here is covered by forest. The forests of the Nature Reserve are Colchian type and there are nearly 60 tree species (Gamkrelidze et al. 2019).

In 1933, dinosaur footprints were found in Sataplia, unfortunately, only 196 footprints of dinosaurs are preserved today, although originally there were more 250 footprints

there. The dinosaur footprints of Sataplia imprinted into the Lower Cretaceous limestones. These imprints are unique because there are two stratigraphic levels: lower imprinting of the predatory dinosaurs and at 1.7 m above it—of herbivorous ones (Fig. 4.1) (Gamkrelidze et al. 2019).

400 m from Sataplia dinosaur footprints there is the entrance of Sataplia karst cave. It was first noticed in 1925 by Kutaisi state museum employee P. Chabukiani, who was able to obtain protection by local government for this valuable monument.

The length of Sataplia cave is 600 m, average height −10 m and average width −12 m. 100 m from the entrance of the cave, there is a cupola-shaped hall (maximum height —6 m), where the beauty of stalactites and stalagmites astonishes visitors (Fig. 4.2). The air and water temperatures in the cave are nearly equal 12–13 °C during all seasons of the year (Gamkrelidze et al. 2019).

From Sataplia nature reserve in 15 km distance is located Kumistavi (Prometheus) karst cave. Kumistavi Cave is one of Georgias's natural wonder monuments with breathtaking example of stalagmites, stalactites, underground rivers, lakes and "petrified waterfalls" (Fig. 4.3). Total length of Kumistavi cave is about 11 km, of which 1600 m is open to visitors, from which 280 m is a karst lake. Totally the cave has 22 halls, with their maximum height of 21 m, of which six are currently open to tourists. Air temperature in the cave is 15–17 °C and water temperature—13 to 14 °C.

The cave was discovered and studied by Georgian speleologists (team leader J. Jishkariani) in the early 80's of the twentieth century. It is part of a large cave system, united by one underground river. Currently, about 30 km of the river is investigated, which is about half the length of the entire cave system. In 1985 the conversion of the cave into a sightseeing tourist destination began.

Today Kumistavi karst cave is one of the most tourist spots of Georgia. It is a unique experience to feel the underworld atmosphere in all its glory and colors. Tourists may choose between walking tours along route, and the boat tour along the

Fig. 4.1 Sataplia footprint of dinosaurs imprinted into Lower Cretaceous limestones: **a**-of predatory dinosaurs, **b**-of herbivorous dinosaurs (Gamkrelidze et al. 2019)

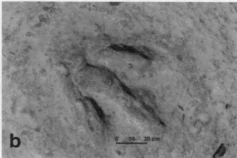

Fig. 4.2 The cupola-shaped hall of the Sataplia karst cave (photography by David Mirzashvili)

Fig. 4.3 One of halls of Kumistavi karst cave

underground river. Visitors are amazed with breathtaking views of stalactites, stalagmites, "petrified waterfalls", underground rivers and lakes of the cave (Gamkrelidze et al. 2019).

Kumistavi owes a famous Caucasian myth about Amirani for the sonorous naming of the "Cave of Prometheus". The legend says that Amirani, like Prometheus, angered the gods and was punished. Days and nights eagle tortured him by eating his liver; however, in contrast to the Greek giant, cruel gods chained Amirani not to the rock, but somewhere inside a huge cave presumably in Kumistavi (Gamkrelidze et al. 2019).

As mentioned above, 8 km down from State Natural Reserve of Sataplia, and 24 km from Kumistavi cave there is a worldwide known SPA resort—Tskaltubo. The resort is famous for its unique radon-carbonate mineral springs. The natural temperature of the water that varies between 33 and 34 °C enables it to be used without preliminary heating.

4.2 Tskaltubo Resort-Town and Mineral Water Deposit

Tskaltubo mineral water flows in the form of numerous sources 12 km north-west of the city of Kutaisi, in the valley of the karst Tskaltubo river 98 m above the sea-level. The ascending water sources are confined to the limestones and dolomites of the Lower Cretaceous, which lie on the variegated, water-resistant suite of the Upper Jurassic. They themselves overlap from above with the water-resistant marl-clay deposits of the Aptian and Albian, contact with which concentrated mineral water outlets (Tsertsvadze et al. 1970; Tsertsvadze 2017).

On the area of outcrops of mineral water, alluvial-proluvial sediments are widely developed, represented in the lower part mainly by magnetite quartz-arcos sand, and in the upper part by—clay and loam.

Mineral water, rising from a depth of 500–600 m along the cracks of bedrock, then falls into the mentioned sands. Their water is under pressure due to the presence of waterproof clay cover. In those places where the latter is less resistant to mechanical composition or to erosion conditions, water comes out in the form of numerous griffins at an average height of 92.5–93.5 m above sea level (Tsertsvadze et al. 1970).

The combination of outlets creates an ellipse-like area with a long axis of about 600 m. It is called the balneological area. The total flow rate of Tskaltubo sources is 250 lt/s. In the annual mode, the value of the total production rate of sources fluctuates noticeably, which is associated with changes in atmospheric precipitation in the field of nutrition and in the field of discharge.

Tskaltubo mineral waters are famous for their stable physical and chemical composition and they are categorized as slight radon chloride–magnesium waters.

In the gas composition of Tskaltubo water, the first place (96–97%) is nitrogen, rare gases participate in the amount of up to 1.4% and about 2% is carbon dioxide and oxygen.

A characteristic feature of the water is the radioactivity. With an increase in flow rate, the radioactivity of waters decreases, and vice versa, but the radioactivity of waters obtained directly from bedrock, with an increase in flow rate increases.

Tskaltubo Mineral water cures the following diseases: arthritis, peripheral nervous system, cardiac and vascular, skin, disbalance, endocrine system, gynecological issues, rheumatism, poliomyelitis, cerebral palsy. Waters are recommended for adults and children (over 3) as well (Fig. 4.4).

In 2015 the new medical thermal center Tskaltubo-BE Healthy was established, equipped with the latest technologies and modern infrastructures. The center is notable for the wide range of services offered in its calm and comfortable

Fig. 4.4 Tskaltubo—Bathhouse №7 entrance area Copyright © 2014 Daniel-tbs

environment. It offers treatments with healing waters and various health procedures. For example: Skin—restorative procedures, Body Fitness, Spa and Wellness body massage, Wellness hydrotherapy procedures, Application of mineral mud (Peleoidoterapy), Diode laser, Gym and many other health-improving practices.

4.3 Borjomi Resort-Town and Mineral Water Deposit

Borjomi is a resort town situated in the picturesque Borjomi Gorge on the eastern edge of the Borjomi-Kharagauli National park, in the central part of the Adjara-Trialeti folded zone. The town is noted for its mineral water industry (which is the number one export of Georgia).

The geological structure of the central part of the Adjara-Trialeti zone, to which the Borjomi mineral water is confined, is characterized by the presence of a stratigraphic section from the Albian to Pliocene.

The oldest formations are represented by a thick volcanic-sedimentary sequence of the Albian and Lower Turonian, which is distributed in separate exposures along the northern slope and ridge of the Trialeti Range (see Fig. 3.32).

The Upper Turonian-Maastrichtian deposits are entirely carbonate rocks. A rather large exposure of limestones and marls of this age is present in the core of the so-called Lomismta anticline, where they are conformable replaced by similar rocks of the Danian and Lower Paleocene.

Paleogene deposits are represented in two facies—in the sandstone-siltstone flysch, so-called Borjomi flysch, and shallow clayey-marly deposits in the areas of the ancient cordillera along the edges of the Adjara-Trialeti zone. Flysch deposits with a thickness of 1500–200 m are exposed in the axial part of the zone in the form of two fairly wide strips in the arched parts of the Borjomi and Lomismta anticlines (Fig. 3.32).

The next Middle Eocene volcanic-sedimentary formation occupies the largest area of the central part of the Adjara-Trileti zone and is composed of fine-grained tuffs of subalkaline basalts (Likani suite), subalkaline basalts, trachytes, dacites and andesites (Kvabiskhevi suite), and rudaceous heavy-bedded tuff breccias, lava breccias and lava sheets of low-titanium basalts (Dviri suite) with a thickness of up to 1700–1800 m.

The Upper Eocene terrigenous sediments, with a thickness of 600–1000 m, lie everywhere transgressively and are preserved only in synclines.

Oligocene—Lower Miocene formations of the so-called Maykop series were preserved in the same areas as the Upper Cretaceous deposits. They are mainly represented by clay-sandy rocks with a thickness of 600–1500 m.

The Upper Miocene-Lower Pliocene thick volcanogenic-effusive formations are developed in the southern part of the Adjara-Trialeti zone and are known as the Goderdzi and Kisatibi suites.

In the central part of the Adjara-Trialeti zone, several Quaternary volcanic centers of andesite composition are located.

The tectonic structure of this part of the zone is characterized by the existence of an axial uplift, on both sides of which fairly long folds of the first order are developed, which replace each other echelon-like along the strike and are complicated by reversed faults with upthrown northern sides.

Borjomi is a balneological and climatic resort. According to the relief and microclimatic peculiarities, there are several districts in Borjomi, of which Likani, porridge and plateau are distinguished by particularly favorable conditions for treatment. The main treatment of the resort is mineral water, which is used for drinking and balneological procedures (Fig. 4.5). Borjomi water is mined from 9 production wells with a depth of 200–1500 m with a natural temperature of 38–40 °C for thermal water. Mineralization of drinking water is 5.0–7.5 g/lt.

Other medical factors of Borjomi are climatotherapy, mud treatment and others. The best time for climatotherapy is June–October. Medical indications: Chronic diseases of the gastrointestinal tract, liver and bile ducts, metabolic disorders, some cardiovascular diseases. The season lasts all year round.

The natural exits of carbonic acid bicarbonate sodium mineral water are located in the gorge of r. Borjomula and are confined to the axial part of the Borjomi anticline, which within Borjomi is composed of the sandstone-siltstone flysch (Borjomi flysch) of the Paleocene-Lower Eocene and intersected by three erosional depressions by the gorges of the Gujaretistskali, Borjomula and Kura rivers.

There are many different views on the genesis of Borjomi mineral water, which generally come to the following: (1) Borjomi water is ancient (buried) water; (2) It is formed as a result of exchange-absorbtion processes between infiltration water and the Upper Cretaceous complex of carbonate and (or) Paleogene flysch rocks; (3) Water enters the Upper Cretaceous and Paleogene sediments from the Cretaceous volcanogenic suite and, possibly, from older rocks in an already formed form; (4) Borjomi water is formed as a result of mixing highly mineralized soda water, coming from volcanogenic Cretaceous, with fresh infiltration water; (5) Carbon dioxide, according to all researchers, has a deep origin and is associated either with post-volcanic processes or with zones of metamorphism. According to Tsertsvadze (2017) there is an active discharge of the deep CO_2, above the deep faults, along with CO_2 and microelements, fluids and soda are carried out.

Fig. 4.5 Borjomi Central Park, Main Spring of mineral water (photography by Zviad Avaliani)

Along with this, based on the study of stratigraphy, lithology, surface and deep structure of the Borjomi region, as well as careful geological and hydrogeological analysis of the primary material on boreholes, it was suggested (Gamkrelidze and Lobjanidze 1984) that the Borjomi mineral water is of stratified type and related to the Borjomula carbonaceous subsuite of the Lower Paleocene. However, Borjomi water has a limited areal distribution due to the location of the area of feeding and discharge of the water-bearing horizon.

Another potential Geopark—Vardzia Upper Miocene megacaldera and ancient Vardzia rock-cut town is described in detail in the spetial works of Okrostsvaridze et al. (2016, 2019) and therefore is not considered here.

4.4 Kazbegi Quaternary Volcano and Keli Volcanic Highland

Kazbegi is the northernmost beautiful district of Georgia, which is located on the Main range of the Greater Caucasus, approximately at 130 km from Tbilisi (Fig. 4.6). Stepantsminda town is the administrative center of the district. Historically Kazbegi districts is called Khcvi province (Ravine province—in Georgian). Area of the district is 1082 km^2 where up to 4000 residents live. Despite of its very small territory, due to its interesting geological structure, fascinating landscape and long history, the Kazbegi district is considered one of the most promising potential geoparks in

Fig. 4.6 General view of the Kazbegi district from the South. In the foreground-village of Gergeti, in the central part-Sameba Church, in the foreground-Kazbegi volcano

Georgia (Gamkrelidze et al. 2011, 2019). Added to that is that around the Mount Kazbegi since 1979 has been existed the Kazbegi National Park, with a total area exceeding 9000 ha. The area includes beech forests, subalpine forests and alpine meadows. Many of the plants and animals in the reserve are endemic to the Caucasus region.

In the small area of Kazbegi district long geological history (including Paleozoic to Pleistocene) and completely different genetic type rocks outcrop. Here crop out Paleozoic granitoid (Dariali massif), Meso-Cenozoic sedimentary rocks, Pleistocene subaerial volcanic centers and lava flows, mostly of andesitic composition. The most prominent geological formation of the region is the Kazbegi Pleistocene volcanic structure (Fig. 4.7) (Gamkrelidze et al. 2019).

Stratovolcano Kazbegi (5053 m) is a dormant construction of the district and one of the major summits of the Greater Caucasus, located on the border of Georgia and Russian Federation (Fig. 4.7). It is the seventh-highest summit in the Caucasus Mountains. Kazbegi is also the second-highest volcanic summit in the Caucasus, after the volcano of the Elbrus. Mount Kazbegi is covered with everlasting snow and glacier and with reference to its beauty is named 'bride' of the Greater Caucasus. The first climb to the summit of Kazbegi was made by English mountaineers Freshfield, D. W., Moore A. W. and Tucker C. in 1868. The total area of Kazbegi's glaciers is 135 km^2. The best-known glacier is Devdoraki, which creeps down the north-eastern

slope to a gorge of the same name, reaching a level of 2295 m (Gamkrelidze et al. 2019).

The geological structure of the Kazbegi volcano is quite complex and several stages of its formation are distinguished: Early Pleistocene, Middle Pleistocene and Late Pleistocene (Skhirtladze 1958). It is noteworthy that the Goristsikhe andesitic flow of this volcano is dated by the radiocarbon method as 6000 ka (Burchuladze et al. 1976). However, the U-Pb dating of zircons separated from the same flow, at the isotope laboratory of the National University of Taiwan showed the age of 32,000 ± 1.6 years (La-ICP-MS study) (unpublished data), which makes the Holocene age of the Kazbegi volcano controversial (Gamkrelidze et al. 2019).

At the first stage of its action the Kazbegi volcano mainly erupted lava of andesitic composition with small amount of pyroclastic material. A small amount of dacitic pyroclastic material is also observed at this stage. The activity stage of volcano is dated by K-Ar method as 455,000 ± 40 Ma (Chernishev et al. 1999).

The Kazbegi volcano was the most active at the second—Middle Pleistocene stage. At this time, the main structure of the volcano formed. The volcanic rocks of this stage consist mainly of site flows and their pyroclastic equivalents, with minor dacites and basaltic bodies. The K-Ar isotopic age of these rocks ranges from 235,000 ± 40 to 185,000 ± 30 Ma (Chernishev et al. 1999).

Fig. 4.7 Kazbegi volcano view from the South-Eeast. In the foreground—Gergeti andesitic flow

The third, final stage of the volcano Kazbegi activity was in late Pleistocene. At that time, only a small amount of pyroclastic material of the andesitic composition was erupted and flows of the same composition were spread. The volcanic rocks of this activity stage are dated by the K-Ar method as 50,000 ± 20 Ma (Chernishev et al. 1999). As it was noted above, the zircons dating by U-Pb method, from the same period volcanic flow of Goristsikhe, showed the age of 32,000 ± 1.6 million years (La-ICP-MS study).

Apart from the volcano Kazbegi, in the district other—Pleistocene volcanoes were also active in the area, namely the Kabarjina, Sakokhe, Sadzele, and others, which had small scale, although they are also interesting sites for geotourism.

Intense subaerial volcanic processes also occurred in the southern part of the Kazbegi volcanic region during the Quaternary Period, which resulted in the formation of a volcanic plateau on Meso-Cenozoic sediments, known in the geological literature as the Keli volcanic plateau. There are about 30 extinct volcanic centers on this plateau. In volcanic centers are located lakes, creating beautiful landscapes (Fig. 4.8). The Keli volcanic plateau is a remarkable geotourism site, but unfortunately the western part of this plateau has been occupied by Russia since the 2008 Russian-Georgian war, creating significant communication problems here (Gamkrelidze et al. 2019).

Keli volcanic plateau (20 km × 30 km) is located on the Main Ridge of the Greater Caucasus, western side of Djvari Pass. On the plateau there are thirty volcanic edifices, which consists mostly of monogenetic volcanic structures, of andesitic and trachyandesitic composition (Gamkrelidze et al. 2019).

The rivers Tetri Aragvi (White Aragvi), Patara Liakhvi, Ksani, Didi Liakhvi and Terek originate from the Keli volcanic area. Western Khorisar (elevation 3736 m) and Didi-Nepiskalo (3694 m) are the highest summits. The Patara-Nepiskalo volcano is strongly degraded by glacial action and have had a caldera. Other volcanoes are Sharkhokh, Northern Shadilkhokh, and Southern Shadilkhokh in the western Kaidon range (Gamkrelidze et al. 2019).

Georgian geologist Nikoloz Skhirtladze first studied the volcanoes of the Keli volcanic plateau and divided the volcanism into glacial and postglacial stages (Skhirtladze 1958). Later research indicated three phases of volcanism: A first phase 245–170Ka involved lava dome formation including Kordieritovyi dome, Patara-Nepiskalo volcano and Kabarjina–Sakokhe center and it is associated a lava flow in the Aragvi valley. A second phase 137–70 Ka with large scale effusive activity included the "Pyramidal Peak" Volcano which dammed local rivers with lava flows. From 90 Ka activity formed Didi-Nepiskalo and the Ekisom and Khorisar domes (Gamkrelidze et al. 2019). The third phase less than 30 Ka with evidence that it continued into the Holocene formed Eastern Khorisar (Dzotsenidze 1972).

4.5 Dariali Paleozoic Granitoid Massif

Dariali Paleozoic granitoid massif crops out in the river Dariali gorge at about 7 km in the distance from village of Stepantsminda, where the height of the magmatic rocks granitoid cornices is more than 1000 m, making an unforgettable impression on a viewer (Fig. 4.9). Currently, Dariali

Fig. 4.8 Keli volcanic plateau-view from the South-Eeast. In the foreground volcano Patara-Nepiskalo (photography by Jussi Saha)

Fig. 4.9 The beginning of the Dariali Gorge from South (from Georgia) Copyright © 2008 Kober

Gorge is on the border between Russia and Georgia and it is a part of the Georgian Military Road. The gorge was carved by the river Terek, and is approximately 14 km long. It is noteworthy that the river Dariali is the only river on the main ridge of the Greater Caucasus that starts on the southern slope of the ridge and then turns to the north to 180 degrees, crossing the main ridge of the Greater Caucasus. Such a change in the river flow was probably caused by the formation of a Quaternary Keli volcanic plateau that blocked the river to the south, after which it began to flow northward and cut through the Dariali Paleozoic granitoid massif.

The Dariali pass was historically important as one of only two crossings of the Caucasus mountain range, the other being the Derbent Pass, between Azerbaijan and Dagestan (Gamkrelidze et al. 2019).

Information on the significance of the Dariali pass of the Greater Caucasus can be found in Roman manuscripts as early as BC, and here the ruins of an ancient fortress are still visible. The pass served as a hub point for many roads connecting North and South Caucasus and remained open for traffic for most of its existence. Due to its charm and history, the Dariali pass is one of the most romantic places in the Greater Caucasus Ridge, and poets and artists dedicating remarkable creations to it (Fig. 4.10).

The Dariali pass is cut in the Late Paleozoic granitoid massif, which is cropped out on the North-East of Kazbegi volcano. Paleozoic granitoid massif of Dariali represents the easternmost outcrop of Caucasus crystalline basement. It thins out in a sublatitudinal fashion and has length of more than 7 km, while width reaches 4 km. Much smaller Gveleti massif, having similar age and genesis, crops out 1500 m

southwards. Between these two, Lower Jurassic black shales, sandstones and quartzites are cropping out and known as Kistinka suite (Fig. 4.11). The massifs and Kistinka suite olso are cut by numerous diabase dikes, mainly of

Fig. 4.10 Dariali pass—Painting by Ukrainian artist. Rufin Sudkovsky (1884 year)

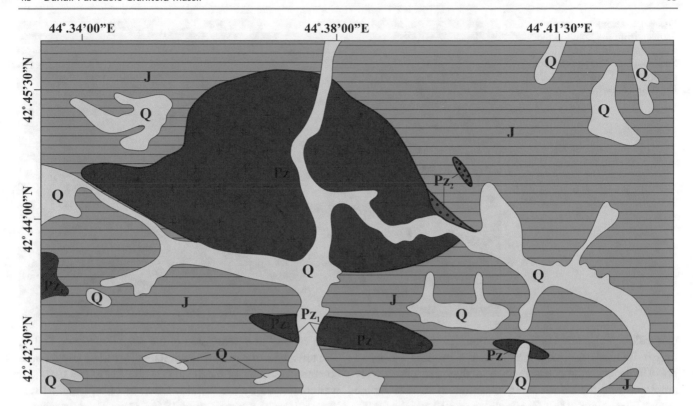

Fig. 4.11 Schematic geological map of the Dariali and Gveleti granitoid massifs (Gamkrelidze and Shengalia 2005). 1—Quaternary deposits, 2—Jurassic deposits, 3—Lower Paleozoic milonitized crysialline schists, 4—Middle Paleozoic metamorphites, 5—Upper Paleozoic granitoids, 6—Protrusive contacts

Jurassic age, exceeding 100 in number and ranging between 0.3 and 12 m in thicknesses (Gamkrelidze and Shengelia 2005).

Crystalline rocks forming Dariali and Gveleti massifs are predominantly granitoids broken down and milonitized in Alpine time (Gamkrelidze and Shengelia 2005). Isotopic dating of these granitoids was carried out. Muscovite from pegmatite turned out to be 321 ± 6 Ma (Middle Carboniferous) (Dudauri et al. 2000). The zircons U-Pb ages of Dariali massif granodiorites are determined as 314 ± 1.7 (unpublished data), in Taiwan National University, using La-ICP-MS study.

Quartz-plagioclase-hornblende and pegmatoid veins, often found in Dariali massif sharply differ from each other in mineral composition and genesis and clearly are of different ages. In particular, the first post-Paleozoic, most likely Middle Jurassic, and the second, Variscan (Gamkrelidze and Shengelia 2005).

Strong tectonization of Dariali granitoids, as a result of which granitoids are compressed, broken down and milonitized, creating a shaly texture, especially characterizing contact zones.

The thickness of contact, strongly milonitized zone usually ranges from 50 to 150 m (Gamkrelidze and Shengelia 2005). The structure of mylonitized granitoids is variable. Meet fine-milonite, coarse-milonite, coarse-blastomilonite, brecciated and other structures. In general, the rock is characterized the structure that most responds to blastomilonite (Gamkrelidze and Shengelia 2005). In the contact area the Dariali and Gveleti massifs with the Lower Jurassic Kistinka suite are in complete concordant orientations in the space of schistosity and cleavage in the Kistinka suite with schistosity and blastomilonite banding in granitoids. In some places, a complete coincidence of the orientation of the foliation (schistosity) of apogranite milonites and the axial cleavage in the Jurassic deposits is observed. In places, these textures are jointly deformed into gentle folds with subhorizontal axial surfaces and apexes. Partial milonitization and often the budding of Middle Jurassic diabase dikes are also noted (Gamkrelidze and Shengelia 2005).

All this indisputably testifies to joint, alpine in age, tectonization granitoids and Lower Jurassic Kistinka suite, often obscuring traces of movement along contact of these rocks.

To the contact zone of the Dariali and Gveleti massifs, they are associated with their movement together with fragments of the Paleozoic cover, inside the shaly stratum of the Jurassic in the Late Alpine time.

Taking into account the above, as well as a sharp submersion to the east of the Ardon River of the crystalline core of the Greater Caucasus and its overlapping by powerful shaly strata of the Lower Jurassic, it seems to us quite natural

Fig. 4.12 Orthodox church of Sameba from the South in the background part of the Southslope of Kazbegi volcanic construction Copyright © 2018 Hotel Sno Kazbegi

that Paleozoic granitoids and metamorphites of the Dariali massif in modern structure represent a rootless blocks of pre-Alpine crystalline basement and they are protrusively emplaced into the Lower Jurassic sediments in Late Alpine time (Giorgobiani 2000; Gamkrelidze and Shengelia 2005).

It should be noted that the massif is characterized also by a large number of fine grained spherical mafic inclusions of plagioclase and quartz composition. Partial melting and hybridism represented the magma generation mechanism of these rocks (Okrostsvaridze and Tormay 2013).

In Kazbegi district near the geological site are many ancient historical monuments as well. Among them is distinguished magnificent orthodox church of Sameba (Holy Trinity) near village Gergeti. It is constructed inthe fourteenthcentury, probably by the Georgian King Giorgi V the Brilliant. Gergeti Samebais situated at 2170 m a.s.l. on the andesite volcanic flow of Kazbegi volcano. In eighteenth century, the church turned into storage for main Georgian relics that were transported here in the time of Persian invasion to Tbilisi. In the beginning of the twentieth century, Soviet government had closed the church, and it was returned back to the Georgian Orthodox Church only in the 1990's. Currently, Gergeti Church of Sameba is very popular touristic destination of Georgia (Fig. 4.12).

4.6 Dmanisi Hominids Site and the Mashavera Gorge Basaltic Flow

Dmanisi is a town which is located at 90 km southwest of Tbilisi on the Quaternary Javakheti volcanic plateau, in the Mashavera river valley. This area had been settled since the Early Bronze Age, though Dmanisi is first mentioned in the

ninth century as a possession of the Arab Emirate of Tbilisi. The town was of particular importance, growing into a major commercial center of medieval Georgia, bat in the 1080's, it was conquered by the Seljuks. Later, in 1124, Dmanisi was liberated by the king David IV of Georgia. After that it became a royal fortress and was an important economic and strategic town. At the end of the fourteenth century, the Turco-Mongol armies of Tamerlane laid waste to the town. After this Dmanisi never recovered and declined to a sparsely inhabited village by the eighteenth century. Dmanisi district offers two important geosites: Dmanisi hominid site and the river Mashavera Quaternary basalt flow. Dmanisi hominid site discovered in 1991, which is situated on the hill, elevated 80 m above the confluence of the Mashavera and Pinezauri rivers (Fig. 4.13). It takes over 300 m^2 and it is built up by Mashavera river Quaternary basalt flow. Five

Fig. 4.13 Dmanisi hominid site, after the archaeological works

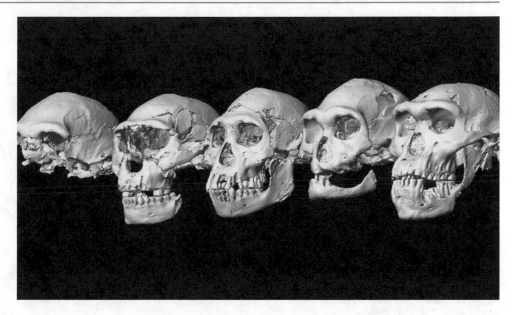

skulls, four mandibles and many other bones of *Homo Georgicus* have been discovered in this basalt flow (Gabunia and Vekua 1995) (Fig. 4.14). It was the earliest known evidence of hominins outside Africa before stone tools dated to 2.1 million years were discovered in 2018 in Shangchen, China (Zhaoyu et al. 2018).

The Dmanisi hominid specimens are the most primitive and small-brained humans found outside the Africa continent, and they were attributed to *Homo erectus* sensu lato. Lately, they were identified as a subspecies *Homo Erectus Georgicus* or even *Homo Georgicus* representing transitional stage between *Homo Habiles* and *Homo Erectus* (Gabunia et al. 2000). It is widely recognized that the Dmanisi discoveries have changed the knowledge concerning the migration of Homo from Africa to the European continent.

It is well known that the limit for effective dating with isotopic carbon method (C^{14}) does not exceed 30,000 years, which makes it impossible to directly determine the age of the hominids of Dmanisi. Because of this, their dating was only possible with the use of other isotope pairs and through geological processes analysis.

The hominids and faunal remains are placed above basalt flow, which by the $^{40}Ar/^{39}Ar$ method are dated as 1.85 ± 0.01 Ma. All the remains are covered with 1 m thick andesitic pyroclastics (tephra), which are eroded and overlying fluvial sediments. Volcanics by the $^{40}Ar/^{39}Ar$ method are dated as 1.81 ± 0.05 Ma (Lumley et al. 2008). As a result of the data analysis, it is clear that Dmanisi hominid died between these ages. Among the remains 5 hominids, 4 are between 13 and 40 years old, whereas 1 hominid is older because it does not have teeth. Such age variations of Dmanisi hominids and geological conditions of fossilization suggest that they died during volcanic eruption from toxic volcanic gases and later were covered with pyroclastic material (Lumley et al. 2008).

The temporal link between the mass extinction and large igneous provinces is well known. It is notable that the best-constrained examples of death-by-volcanism record the main extinction pulse at the onset of (often explosive) volcanism, suggesting that the rapid injection of vast quantities of volcanic gas (CO_2 and SO_2) was the trigger for a truly major biotic catastrophe (Bond and Wignall 2014). Based on the results of our work, we consider that the extinction of many vertebrates living in the area of the Javakheti volcanic plateau and habitats of the Dmanisi Paleolithic site among them, might have been caused by mass extinction from gases released as a result of powerful volcanic eruptions (Okrostsvaridze 2016).

According to the analysis of geological processes in the region during the Quaternary period, the massive death of Dmanisi hominids was probably caused by the eruption of the Emlikli volcano (3054.8 m) on the Javakheti plateau. It lies about 20 km from the site of the hominid burial, and the rocks of this volcano reveal a complete petrographic and chemical identity with the composition of the tephra that covers the remains of the Dmanisi hominids. According to modern $^{40}Ar/^{39}Ar$ isotopic dating data, this volcano has been active in the interval from 1.9 million to 1.23 million years ago (Meliksetian 2018). So based on these data, one of the first major eruptions of Volcano Emlikli would probably have been the cause of the death of Dmanisi hominids. However, if we share this view, then there must be other remains of buried hominids in the region that have not yet been discovered (Gamkrelidze et al. 2019).

Late Cenozoic volcanic plateau of Javakheti, where Dmanisi is located, geologically is also a very interesting

Fig. 4.15 Part of Javakheti volcanic plateau in spring. In the foreground—Paravani Lake, and in the background-Pleistocene generation the Abul-Samsar Volcanic Ridge

object. Late Cenozoic collisional volcanic activity, between the Arabian plate and the Eurasian Continent shaped vast volcanic highlands, present in the Georgian, Turkish and Armenian neighboring territories. The highland, located in Georgia, is known as the Samtskhe-Javakheti volcanic highland ($\sim$4500 km^2), which is divided into two parts by the Mtkvari (Kura) river canyon: western—Samtskhe and eastern—Javakheti. Javakheti Plateau is alpine steppe, with numerous wetlands and wonderful lakes (Fig. 4.15). It is crossed from south to north by the Javakheti and Abul-Samsari volcanic ranges, where more than 20 volcanic cones are observed (Gamkrelidze et al. 2019).

As mentioned above, the Mashavera river Quaternary basaltic flow comes from Javakheti volcanic plateau and runs along the river gorge for 20 km. The basaltic flow crop out very well in the relief and represents an important geosite of the region. Its thickness varies between 15 and 17 m and overlies alluvial deposits of the river gorge. The basalt flow builds left bank of the river and makes excellent geologic outcrop, where contact of high temperature lava and underlying alluvium can be observed (Fig. 4.16). It should be noted that, lots of faunistic material are found along the surface of the flow, as are teeth and jawbones of elephant, wolf, deer, ox and horse. That's another important evidence of massive extinction in the region, caused by volcanic activity (Gamkrelidze et al. 2019).

As we mentioned in the introduction to this chapter, the Dmanisi district is the oldest, historic part of Georgia (Kvemo Kartli region), where with the natural site are many ancient historical monuments as well. Near modern Dmanisi are ruins of Dmanisi fortress city of sixth—fifteenth centuries, where old houses, streets, wine cellars and other

Fig. 4.16 Mashavera river Quaternary basaltic flow. Under the flow alluvial sediments of the river can be seen

medieval buildings are well preserved. Dmanisi Castle was built on the highest point of the town and represented its fortress. The castle owned by ancient Georgian family of Orbeliani was in use until the end of the eighteenth century. Currently its ruins are still standing and it is an important tourist destination. It should be noted that in the heart of the Dmanisi historic site, a sixth century Orthodox Christian church, Dmanisi Sioni is located (Fig. 4.17). It is protected by the state as an important Cultural Monument of National Significance (Gamkrelidze et al. 2019).

Fig. 4.17 A sixth century Orthodox Christian church of the Dmanisi Sioni (photography by Lasha Gabelia)

Thus, in Dmanisi town are put together Dmanisi hominid site of 1.81 Ma, Mashavera river Quaternary basaltic flow, ruins of Dmanisi medieval Castle and a sixth century Orthodox Christian church. Bringing together these unique natural and cultural objects and their proximity to Tbilisi are good preconditions for Dmansi geopark establishment.

4.7 Dashbashi Canyon

Dashbashi Canyon (Fig. 4.18) is located 110 km from Tbilisi and 2 km from town of Tsalka, near the village Dashbashi. It is formed on Javakheti Quaternary volcanic plateau, in Khrami river gorge, on the 1100–1500 m. Dashbashi Canyon is one of the most interesting and fascinating natural monuments in Georgia (Fig. 4.19). On the eastern slopes of the canyon there are numerous waterfalls and natural as well as man-made caves. Because of that Dashbashi Canyon currently is one of the most popular tourist destinations in Georgia. Geologically this canyon is interesting also, because the basaltic flows of Javakheti volcanic plateau crops out very well here. The thickness of each flow varies within 20–30 m, here 6–7 flow streams, with a total thickness of 120–240 m (Fig. 4.19).

Javakheti volcanic plateau is an eastern block of the Late Cenozoic Samtskhe-Javakheti volcanic province of the Lesser Caucasus. It is bordered on the western (Samtskhe) block by an active regional fault. Over 7.5 Ma in comparison with the western block went down about 130 m (Okrostsvaridze et al. 2016). During this period Javakheti block is covered with thick basaltic (dolerite) flow, so-called Akhalkalaki suite (Skhirtladze 1958). As a result of field, petrologic petrochemical and isotopic works it was established that these formations are typical continental flood basalts (traps) (White and McKenzie 1989).

The Javakheti volcanic plateau was formed on Paleozoic continental crust, of the Artvini-Bolnisi block. Here, the micro-plateaus of Akhalkalaki, Tsalka, Gomareti, and Dmanisi are distinguishable, in which basaltic flows are fully identical petrographically and geochemically.

The flows are grey, fully crystalline, coarse-grained, and weakly differentiated massive rocks, which mainly consist of olivine, basic labradorite, monoclinic pyroxene, and titanoaugite. By petrochemical data, they are more related to mid oceanic ridge basalts than to islands arcs. Content of SiO_2 in these flows varies in the range of 49–51%, and that of MgO varies within 6–8%. The $^{143}Nd/^{144}Nd$ parameter varies in the range of +0.51703 to +0.52304, and the

Fig. 4.18 An overall view of continental flood basalt of the Dashbashi canyon (Upper reaches of the Khrami river) (photography by Lasha Gabelia)

Fig. 4.19 Fragment of Dashbashi canyon—"paradise on the Earth". In this picture, three basaltic flows are well distinguished (photography by Giorgi Shermazanashvili)

87Sr/88Sr parameter ranges from 0.7036 to 0.7042 (Okrostsvaridze et al. 2016). Magmatic zircons of the Javakheti plateau basalts have been dated by the U-Pb method in Taiwan National University. The results obtained vary in the range of 2.4–1.6 Ma (Chang et al. 2013).

4.8 Uplistsikhe Rock-Cut Town and Kvakhvreli Cave Complex

Uplistsikhe rock-cut town is located on the left bank of the Mtkvari (Kura) River. The nearest settlement is the village of Kvakhvreli, which is located on the right bank of the r. Mtkvari at a distance of 1.4 km from the complex. The main morphological element of the relief is the Kvernaki meridional ridge (with absolute heights ranging from 700 to 1100 m. It is built up of Miocene terrigenous deposits.

The Kvernaki ridge is developed on the northern limb of the Gori-Kaspi anticline and formed as a result of the erosive action of the r. Mtkvari. On the right bank of the r. Mtkvari in the form of an "erosive witness" is observed a small hill. It

is within this hill in the sandstones of the lower Miocene (Sakaraulo regiostage) that the Kvakhvreli caves are carved.

The Kvakhvreli cave complex includes a bell tower, corridors, a church and dwellings (Fig. 4.20). Kvakhvreli cave in historical sources was first mentioned in the eleventh century, during the struggle for the unification of feudal Georgia. It is also known that in the ancient era it was an agricultural suburb of Uplistsikhe (Dolidze et al. 1990).

Uplistsikhe is one of the oldest settlements in the Caucasus (Fig. 4.21). In written historical sources it was first mentioned in the seventh century. Although archaeological and architectural monuments preserved in its surroundings indicate that the development of sandstones began in the first half of the 1st millennium BC (Kipiani 2002). During the fourth–first centuries BC, Uplistsikhe was a significant religious, political and cultural center. Starting from the fourth century AD, Uplistsikhe gradually loses its strategic importance, and in the thirteenth century, after the Mongol invasion, it generally loses this function.

The Lower Miocene deposits, in which were carved Uplistsikhe rock-cut town and Kvakhvreli cave complex, are

General plan: I. Dark corridor;
II. Corridor;
III. Open corridor;
A. Bell tower;
B. Church;
C. Dwelling with a window;
D. Incomplete dwelling;
E. Auxiliary dwelling.

Fig. 4.20 Church and entrance into the Kvakhvreli complex (photography by Lasha Zubiashvili; General plan according to Dolidze et al. 1990)

Fig. 4.21 Uplistsikhe rock-cut town (photography by Giorgi Shermazanashvili)

lithologically represented by coarse-grained sandstones, gravelites and micro-conglomerates with many quartz fragments. Part of the section is characterized by cross bedding, indicating the formation of sediments in coastal conditions. In addition to river erosion, aeolian bedforms and columns, tables, obelisks, niches, and kettles are observed in sandstones (Fig. 4.22).

Submarine landside folds (Fig. 4.23) are observed in the clay sandstones to the north-east from the ruins of the old village of Uplistsikhe on the border of the Aquitaine (Uplitsikhe) and Sakaraulo regiostages.

4.9   Udabno Upper Miocene Marine and Continental Deposits and David Gareja Monastery Complex

The desert (Udabno) Gareji is located in the extremely southwestern part of the Iori Upland, in the border part of Georgia and Azerbaijan. It is the most waterless and devoid of vegetation area of Georgia. It is composed of marine and continental deposits (Bukhsianidze and Koiava 2018). Here the relief is strongly dissected and characterized by low

Fig. 4.22 Aeolian bedforms (photography by Teimuraz Chelidze)

Fig. 4.23 Submarine landslide folds (photography by Teimuraz Chelidze)

ridges of the north-western direction, which are confined to the outcrops of compact continental sandstones and conglomerates of the Sarmatian, Meotian-Pontian and Akhchagilian ages (Fig. 4.24). At the same time, negative relief forms are associated with Upper Sarmatian so-called Eldari suite, which is represented by continental parti-coloured clays with rare intercalations of sandstones and microconglomerates (Fig. 4.25).

Upper Sarmatian is overlain by The Meotian-Pontian deposits (Shiraki suite) represented by thick continental sandstones with intercalations of clays and micro-conglomerates and rarely volcanic ash.

Structurally, the region represents the south limb of so-called Udabno syncline complicated by reversed fault and inclined to the south anticline (Figs. 4.26). The structure of the so-called Ridge of monasteries (Monastrebis Seri)

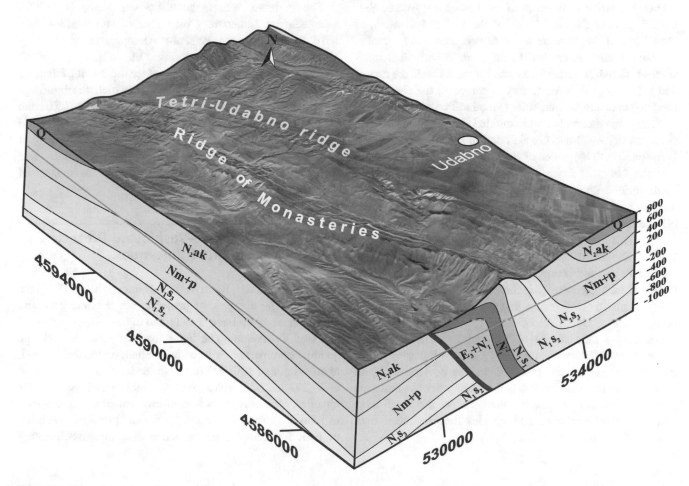

Fig. 4.24 3D model of the relief and geological structure of the Gareja region. Stratigraphical signs: N_2ak—Aghchagilian, $Nm + p$—Meotian and Pontian (Shiraki suite), N_1s_3—Upper Sarmatian (Eldari suite),

N_1s_2—Middle Sarmatian, N_1s_1—Lower Sarmatian, N_1^2—Middle Miocene, $P_3 + N_1^1$—Oligocene- Lower Miocene

Fig. 4.25 Parti-coloured clays of Eldari suite of Gareji desert (Udabno) (photography by Shalva Lezhava)

involves massive sandstones of the Lower and Middle Sarmatian. It is in the sandstones of the Lower Sarmatian, in which one of the monasteries of Gareja monastery complex Natlismcemeli was carved. In the Middle Sarmatian sandstones Chichkhituri monastery and Lavra of David are carved and the Dodo's Rka Monastery is carved in the continental sandstones of the Shiraki suite (Meotian-Pontian).

Gareja cave complex was founded in the first half of the sixth century by David Garejeli one of the thirteen Assyrian Holy Fathers. The center of the Gareja monastery complex was the Lavra of David. Over the time several new cave monasteries added to the complex (Fig. 4.26). Dodo's Rka the first half of the sixth century and the Monastery of Ioane Natlismtsemeli (John the Baptist) the sixth-seventh centuries were built consequently by Dodo and Luciane, disciples of David. The exact date of establishment of Chichkhituri monastery is not known (eleventh-thirteenth centuries). Gareja complex generally includes 19 monasteries.

Batonishvili (1895) notes that the archimandrite sat in the monasteries of David Gareja and Natlismtsemeli while the abbot in the monasteries of Chichkhituri and Bertubani. Since its establishment, the Davit Gareja Monastery Complex has been actively involved in the development and prosperity of the country. It was the place where manuscripts were created and rewritten, the new methods of cave building were developed and besides they had their own school of fresco painting (Fig. 4.27).

The complex was invaded by the Mongols (1265), repeatedly by Tamerlane (Amir Timur), and finally—from 1616 to 1617 by the Persians. Although in 1639, King Teimuraz I again resumed monastic life in it.

The Gareja complex has been under the threat of damage for many years, as the sandstones in which the buildings were carved are subject to intense erosion. In 2007, the UNESCO Gareja Complex was included in the list of World Heritage Sites, and in 2018 by the Pan-European Society of Cultural Heritage "Europa Nostra", this complex was included in the list of seven monuments in danger of destruction.

4.10 Dedoplistskaro-Vashlovani Protected Areas and Mud Volcanoes

Vashlovani Nature Reserve was founded in 1935. Since 2003, a single complex of preserve, the national park and natural monuments have been united in the form of protected areas of Vashlovani. The latter are located in the extreme southwestern part of Georgia, in Kakheti, on the Iori Upland, at an altitude of 100–900 m above sea level and its total area is 35,292 ha. Mud volcanoes are widely developed on this territory. Mud volcano is a landform created by the eruption of mud or slurries, water and gases. The principal condition of their formation is the presence of a thick sedimentary

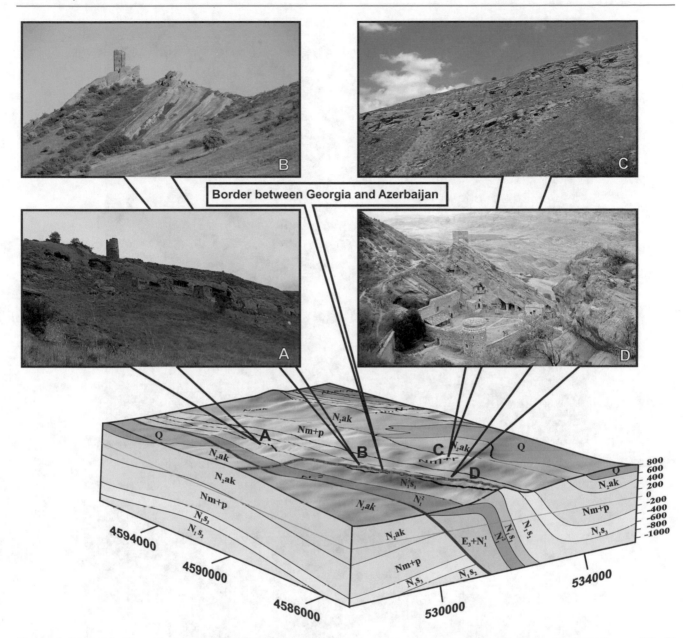

Fig. 4.26 3D model of the geological structure of the Gareja region. A—Natlismcemeli monastery (photography by Bacho Glonti); B—Chichkhituri monastery (photography by Djaba Labadze); C—Dodo's Rka monastery (photography by Shalva Lezhava); D—Lavra of David (photography by Shalva Lezhava). Stratigraphical signs see in Fig. 4.24

cover containing plastic clay members, which represent source formations for mud volcanoes (Limonov 2004).

In Georgia, mud volcanoes distribution area mainly encompasses the Iori Upland and the Gombori range. Minor occurrences are observed in the northern flank foothills of the Trialeti range near the village of Kavtiskhevi (Ebralidze et al. 1975). In total, up to 15 mud volcanoes are observed in the Eastern Georgia (Fig. 4.28).

Fig. 4.27 Icon of the Mother of God in the monastery of Mravaltskaro (photography by Shalva Lezhava)

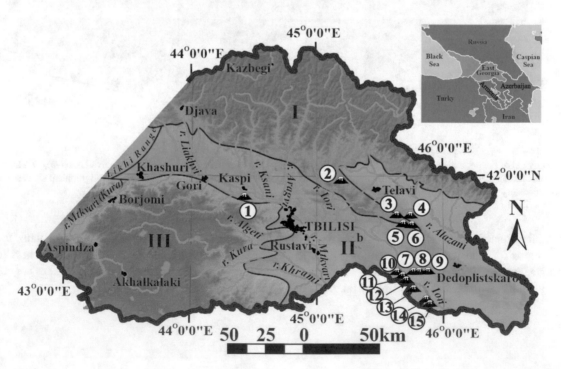

Fig. 4.28 Physical-geographical map of the Eastern Georgia with locations of mud volcanoes (map modified after Ebralidze et al. 1975); I —Greater Caucasus Mountain Chain; II—Iberia (Kura) Lowland; III—Lesser Caucasus Mountain Chain. Mud volcanoes: 1—Kavtiskhevi; 2 —Bakana; 3—Tkhiliskhevi; 4—Akhtala; 5—Lakbe; 6—Pkhoveli; 7—Western Kilakupra; 8—Central Kilakupra; 9—Eastern Kilakupra; 10—Navtis-Chebi; 11—Baida; 12—Aladjigi; 13—Phorphotebi; 14—Northern Tulki-tepe; 15—Takhti-tepe (Southern Tulki-tepe)

Fig. 4.29 Takhti-tepe mud volcano (photography by Bacho Glonti)

Takhti-tepe (SouthernTulki-tepe) mud volcano (Fig. 4.29) has a IUCN category III (natural monument or feature) and is located within the Tulki-tepe anticline, which is built up with Middle and Upper Sarmatian deposits. The volcano is a small hill that is impregnated with 60 active craters, gryphones and salses (satellite craters) of different sizes. The length of the hill is 189 m, the width is 60 m, and it is located at an altitude of 540 m above sea level. The diameter of the largest crater is 3.5–4 m.

NorthernTulki-tepe mud volcano is located near Mount Tulki-tepe in a bowl-shaped synclinal in the Middle Sarmatian deposits of Chobandagi suite, which is represented by brownish-gray, bluish, dark gray and greenish-gray thin clay layers with rare interlayers of light gray marls and tiled brownish sandstones (Fig. 4.30). Its height is 10 m. The maximum diameter of the hill is 250 m, and the height is 480 m. The number of air drains and craters is not more than 20. The diameter of the largest crater is 2.5–3 m.

Phorphotebi mud volcano is located 30 km southwest of Dedoplistskaro (Fig. 4.31). It is listed in the Red Data Book of the Georgian SSR (1982) as a geological and geomorphological monument. Tectonically, it is located on the

Fig. 4.30 Northern Tulki-tepe mud volcano (photography by Bacho Glonti)

Fig. 4.31 Phorphotebi mud volcano (photography by Nini Koiava)

eastern periclinal part of Baida anticline in the form of a hill in the Lower part of Middle Sarmatian deposits called "layers with Criptomactra" and is represented by brownish-gray and bluish clays with rare interlayers of brownish sandstones. The hill has a height 40 m and a diameter 320 m, the number of air craters, gryphones and salses exceeds 80. The diameter of the largest crater is 5 m. The absolute height of the hill is 423 m.

Kilakupra mud volcanoes are located in the central part of Iori Upland and are expressed by three volcanoes:

Western, Central and Eastern Kilakupra (Figs. 4.32 and 4.33). Based on seismic data these volcanoes are related to the dome part of the same name asymmetric anticline structure complicated by faults (Fig. 4.33) and the Middle Sarmatian sediments are supposed as the source of these volcanoes (Glonti et al. 2016).

Western Kilakupra mud volcano is located on the Iori Upland, 25 km southwest of Dedoplistskaro, in the west limb of the Kilakupra anticline. The volcano is an elongated hill 15 m high, 320 m long, 80–100 m wide, with slopes

Fig. 4.32 Kilakupra mud volcanos: **a**—Wester Kilakupra (photography by Bacho Glonti); **b**—Central Kilakupra (photography by Bacho Glonti); **c**—Eastern Kilakupra (photography by Nikoloz Koiava)

dissected by narrow, deep ravines and is composed of the Shiraki suite clays of the Maeotian and Pontian Age. The total number of craters, gryphones and salses is 15. The diameter of the largest crater is 3–3.5 m. The absolute height of the hill is 451 m.

Central Kilakupra mud volcano is located about 1.5 km East away from Western Kilakupra mud volcano, which is stretched from west to east by 120–130 m and, for one glance, is a field completely saturated with oil. The volcano developed in the clays of the Shiraki suite differs from the volcanoes described above by the greatest impregnation of oil in the erupted fluids. On the surface, up to 35 small gryphones are observed of which the largest is 1 m. The absolute height of the volcano is 434 m.

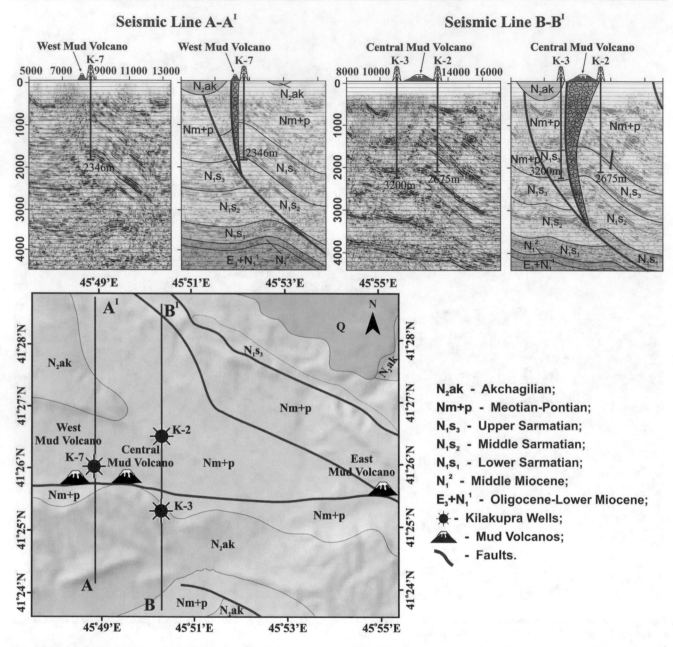

Fig. 4.33 A—Geological map of the Kilakupra area with the location of mud volcanoes and seismic lines; Interpretation of Seismic Line A-A^I and B-B^I (modified after Glonti et al. 2016)

Eastern Kilakupra mud volcano is listed in the Red Data Book of the Georgian SSR (1982) under the name "Second Akhtala (Kilakupra)". It is located 16 km southwest of Dedoplistskaro, in the eastern limb of the Kilakupra anticline, near Mount Akhtala. Volcano is an elongated hill 20–22 m high, whith slopes dissected by narrow and deep ravines. The hill is 340 m long and 190 m wide. Up to 40 craters, griffones and salses are developed in the arch of volcano. The diameter of the largest crater is 2–2.5 m. The absolute height of the volcano is 470 m.

4.11 Summary

The deskribed potential Geoparks of Georgia are: (1) Sataplia dinosaur footprints together with Sataplia and Prometheus caves; (2) Tskaltubo resort-town and mineral water deposit; (3) Borjomi resort-town and mineral water deposit; (4) Kazbegi Quaternary Volcano and Keli volcanic highland; (5) Dariali Paleozoic granitoid massif; (6) Dmanisi hominids site and the Mashavera gorge basaltic flow; (7) Dashbashi canyon; (8) Uplistsikhe rock-cut town and Kvakhvreli cave complex; (9) Udabno—Upper Miocene marine and continental deposits and David Gareja monastery complex; (10) Dedoplistskaro—Vashlovani protected areas and mud volcanoes.

References

Batonishvili V (1895) The geography of Georgia, Book 2, Karti. Published by Mose Janashvili, Tbilisi, pp 1–148. http://dspace.gela.org.ge/bitstream/123456789/6975/1/Sakartvelos%20Geograpia-Kartli%201895.pdf

Bond DPG, Wignall PB (2014) Large igneous provinces and mass extinction: an update. In: Keller G, Kerr AC (eds) Volcanism, impacts, and mass extinctions. Geological Society of America, Special Paper 505, pp 32–41

Bukhsianidze M, Koiava K (2018) Synopsis of the terrestrial vertebrate faunas from the Middle Kura Basin (Eastern Georgia and Western Azerbaijan, South Caucasus). Acta Palaeontol Pol 63(3):441–461. https://doi.org/10.4202/app.00499.2018

Burchuladze A, Janelidze Ch, Togonidze G (1976) Application of radiocarbon method to resolution some of the Pleistocene and Holocene paleogeography of Georgia. The actual problems of contemporary geochronology. Nauka, Moscow, pp 238–243

Chang Y-H, Chung S-L, Okrostsvaridze A, Javakhishvili Z (2013) Petrogenesis of Cenozoic igneous rocks in the Georgian Caucasus. In: Goldschmidt 2013 Conference Abstracts, Florence, Italy, p 853

Chernishev I, Arakelianc M, Lebedev V, Bubnov S, Golcman I (1999) K-Ar geochronology of eruption of the newest volcanic centres of Kazbegi area of the Greater Caucasus. Russian J Earth Sci 1:61–72. https://doi.org/10.2205/1999ES000024,1999

Dolidze V, Kintsurashvili S, Sidamonidze S et al (1990) Discription of historical and cultural monuments of Georgia (Gori, Kaspi, Mtskheta, Kareli, Khashuri regions). Editing Room Georgian Encycl 5:1–504 (in Georgian)

Dudauri OZ, Tsimakuridze GT, Vashakidze GT, Togonidze MG (2000) New data on the age of Dariali massif granitoids. Proc Geol Inst Acad Sci Georgia, New Ser 115:306–310

Dzotsenidze N (1972) Geology of the Volcanic Plateau Keli. Publishing House "Mecniereba", Tbilisi, pp 1–132 (in Georgian)

Ebralidze TP, Bidzinashvili GG, Tatarashvili LI, Geladze SR, Shatirishvili NG (1975) Mud volcanos of Georgia. Funds of VNIGRI, Moscow, pp 1–163 (In Russian)

Gabunia L, Vekua A (1995) A Plio-Pleistocene hominid from Dmanisi, East Georgia, Caucasus. Nature 373:509–512. https://doi.org/10.1038/373509a0

Gabunia L, Vekua A, Lordkipanidze D, Swisher et al (2000) Earliest Pleistocene Hominid Cranial Remains from Dmanisi, Republic of Georgia: taxonomy, geological setting, and age. Science 288(5468):1019–1025. https://doi.org/10.1126/science.288.5468.1019

Gamkrelidze IP, Lobjanidze GE (1984) Geology of the Central Adjara-Trialeti and problem of formation of Borjomi mineral water. Proc Geol Inst Acad Sci Georgia SSR, New Ser 83:1–82 (in Russian)

Gamkrelidze I, Kakabadze M, Okrostsvaridze A (2011) Wide choice of geotraverses and geoparks founding in Georgia. Abstract of the 10th European Geoparks Conference. Langensud, Norway

Gamkrelidze I, Okrostsvaridze A, Maisadze F, Basheleisvili L, Boichenko G, Skhirtladze I (2019) Main features of geological structure and geotourism potential of Georgia, the Caucasus. J Modern Environ Sci Eng 5(5):422–442. https://doi.org/10.15341/mese(2333-2581)/05.05.2019/010

Gamkrelidze IP, Shengelia DM (2005) The Precambrian-Palaeozoic regional metamorphism, magmatism and geodynamics of the Caucasus. Publishing House "Scientific World", Moscow, pp 1–458. (in Russian with extended summary)

Giorgobiani TV (2000) On protrusive origin of granitoid massifs of the Darialy canyon (Greater Caucasus). In: General issues of tectonics. Tectonics of Russia. Moscow, pp 123–125

Glonti VB, Koiava K, Kotulova J, Kvaliashvili L (2016) The structure and geochemistry of the Kila-Kupra Mud Volcano (Georgia). In: Proceedings of the fourth plenary conference of IGCP 610—From the Caspian to Mediterranean: environmental change and human response during Quaternary, Tbilisi, Georgia, pp 79–80. https://doi.org/10.13140/rg.2.2.10679.32161

Kipiani G (2002) Uplistsikhe. Logos program, Tbilisi, pp 1–174. (in Georgian)

Limonov AF (2004) Mud volcanoes. Soros Educ J (8)1:63–69

Lumley MA, Bardintzeff JM, Bienvenu P, Bilcot JB, Flamenbaum G, Guy C, Jullien M, de Lumley H, Nabot JP, Perrenoud C, Provitina O, Tourasse M (2008) Impact probable du volcanisme sur le décès des Hominidés de Dmanissi. Comptes-Rendus Palevol 7:61–79. https://doi.org/10.1016/j.crpv.2007.09.002

Meliksetian Kh. (2018). The generation of collisional magmas on the example of Quaternary volcanism in the territory of Armenia Abstract. Doctoral Dissertation, Yerevan, p 38 (In Russian)

Okrostsvridze A (2016) Quaternary Continental flood basalts of the Javakheti Volcanic Plateau, Lesser Caucasus: reason for mass extinction? In: Proceedings of the fourth plenary conference of IGCP 610—From the Caspian to Mediterranean: environmental change and human response during Quaternary, Tbilisi, Georgia, pp 130–131

Okrostsavridze A, Chung S-L, Lin Y-C, Skhirtladze I (2019) Geology and Zircon U-Pb Geochronology of the Mtkvari Pyroclastic flow and evaluation of destructive processes affecting Vardzia rock-cut city, Georgia. Quatern Int 540:137–145. https://doi.org/10.1016/j.quaint.2019.03.026

Okrostsvaridze A, Elasvili M, Popkhadze N, Kirkitadze G (2016) New Data on the Geological Structure of the Vardzia Cave City, Georgia. Bull Georgian Nat Acad Sci 10(3):98–105

Okrostsvaridze A, Tormey D (2013) Phanerozoic continental crust evolution of the Inner Caucasian Microplate: The Dzirula massif. Episodes 36(1):31–38. https://doi.org/10.18814/epiiugs/2013/v36i1/005

Red Data Book of the Georgian SSR (1982) In Chief editorial board: Kacharava W (Editor-in-Chief), Ketskhoveli N, Maruashvili L, Kurashvili B (eds). Tbilisi, pp 1–255. (in Georgian)

Skhirtladze N (1958) Postpaleogene effusive volcanism of Georgia. Publishing House "Mecniereba", Tbilisi, pp 1–165. (in Ressian)

Tsertsvadze N (2017) Mineral waters of Georgia. Publishing House "Nekeri", Tbilisi, pp 1–242. (inRussian)

Tsertsvadze N, Chikhelidze S, Iuzbashev D (1970) Mineral waters. In: Hydrogeology of SSSR, T X, Georgian SSR: Publishing House "Nedra", Moscow, pp 259–286. (in Russian)

White RS, McKenzie DP (1989) Magmatism at rift zones: The generation of volcaniccontinental margins and flood basalts. J Geophys Res 94(B6):7685–7729. https://doi.org/10.1029/JB094iB06p07685

Zhaoyu Z, Robin D, Weiwen H et al (2018) Hominin occupation of the Chinese Loess Plateau since about 2.1 million years ago. Nature 559(7715):608–612. https://doi.org/10.1038/s41586-018-0299-4

Printed in the United States
by Baker & Taylor Publisher Services